Ergonomics and Safety in Hand Tool Design

Charles A. Cacha, Ph.D., CPE, CSP

CRC Press is an imprint of the
Taylor & Francis Group, an **informa** business

Contact Editor: Ken McCombs
Project Editor: Ibrey Woodall
Marketing Managers: Barbara Glunn, Jane Lewis, Arline Massey, Jane Stark
Cover design: Dawn Boyd

CRC Press
Taylor & Francis Group
6000 Broken Sound Parkway NW, Suite 300
Boca Raton, FL 33487-2742

First issued in paperback 2019

ISBN-13: 978-1-56670-308-6 (hbk)
ISBN-13: 978-0-367-39991-7 (pbk)

Library of Congress Card Number 98-45441

Library of Congress Cataloging-in-Publication Data

Ergonomics and safety in hand tool design / edited by Charles A.
Cacha.
p. cm.
Includes bibliographical references and index.
ISBN 1-56670-308-5 (alk. paper)
1. Tools — Design and construction. 2. Human engineering.
I. Cacha, Charles A.
TJ1195.E74 1999
621.9′08--dc21 98-45441
CIP

Visit the Taylor & Francis Web site at
http://www.taylorandfrancis.com

and the CRC Press Web site at
http://www.crcpress.com

Preface

The purpose of this book is to provide an eclectic approach to the use and design of hand tools. The question of hand tools does not predominate within the repertory of a safety professional, presumably because hand tools are not usually a prominent causation of trauma and disease. Nevertheless, it behooves the safety professional to prepare him/herself in all areas of the safety purview.

This book is broken down by the various disciplines that might apply to hand tools. Each discipline is approached more broadly than might be considered necessary for understanding hand tools. This broad approach is taken because somewhat more-than-necessary information is better than insufficient information.

Acknowledgment

I wish to acknowledge and state my appreciation to Ms. Janet L. Signo, B.S. for her efforts in helping with the preparation of this book.

Charles A. Cacha, Ph.D, CPE, CSP

Table of Contents

Chapter 4

Chapter 5

Chapter 6

Chapter 7

1 History of Hand Tools

Janet L. Signo and Scott C. Jackson

A tool is defined as "an implement used to modify raw material for use." One of man's most distinctive characteristics is the ability to shape and mold the physical world around him. The use of tools has transformed man from a relatively harmless, subtropical vegetarian to a predatory omnivore. Some animals use sticks and stones to accomplish a task, but the ability to make and use tools is one of the significant differences between man and animal. Human beings create things, which is essential to the development of culture and technology.

Over time, people have learned that certain jobs could be done faster and more efficiently with tools. They learned how to use the raw materials around them to solve problems and complete tasks. A main basis of tool design was specialization, using the right tool for a specific job. The use of specified tools has lead humans to overcome their natural limitations.

The appearance of a tool is influenced by the human body, the materials available, and the tasks to be performed. All tools are extensions of the human body and help increase the speed, power, and accuracy nature has given us.

Tools are an important part of archeological remains. These material remains give us clues for studying ancient cultures. How the tools were made and used reflects the lives of past peoples.

There are several processes used in the evolution of tool making. *Reduction* is the process by which a tool is made by reducing the size of a larger object. For example, removing flakes from a stone to make a scraper or sharpening the end of a stick to form a spear. Simple stone and wood tools made by reduction date back over 2.6 million years ago. Toolmakers also developed tools from bone and antler. Toolmakers were limited by the size, shape and amount of raw materials available to them.

Conjunction is the process by which two or more parts are combined, for example, a tone-tipped spear. Tools found using this process date back 150,000 years ago. Toolmakers were able to exceed the limits of the raw materials. Closely related to conjunction is *linkage,* where discrete and separate parts are used in combination. Developments using this process might include the bow and arrow and the mortar and pestle. Linked parts give the tool user a mechanical advantage and help to increase efficiency.

Two or more parts that perform the same task is a process known as *replication.* This process helps improve effectiveness of the tool, while decreasing the chance that the tool will fail or break. Examples of replication are double-barreled shotguns and spears with multiple barbs.

Raw material transformation has helped move man into the modern age. This process involves changing the molecular structure of the actual raw material, for example, tanning hides or extracting ore. This process has increased the raw material available and increased the toolmaker's options.

HISTORY

The Stone Age Period (10,000 B.C.–4,000 B.C.). The first humans used stones that fit into a person's hand and were used to break up firewood and shape other tools. Wooden handles were later added to make them more comfortable to work with. Toolmakers were limited by the inflexibility and brittleness of the material.

Copper and Bronze Ages (4,000 B.C.–1,000 B.C.). Humans discovered how to use fire to separate copper ore from rock. The molten copper was then poured into clay molds to create devices of various shapes, which led to the first metal tools. About 1,000 years later, tin was added to create a stronger, harder metal. Metal brought about a change in the design and use of tools. Tools were thinner, lighter, and easier to handle. These tools were used throughout the Egyptian empire. Finer and more precise work could be accomplished, and new specialized tools were developed.

Iron Age (1,000 B.C.–400 A.D.). Iron replaced copper and bronze as the main raw material. Iron, which lasts longer, produced sharper edges and made it easier to do precise work. Iron tools of Roman origin closely resemble the tools of today.

Medieval Period (400 A.D.–1500 A.D.). This was a very stagnant period. The tools were similar to those developed by the Romans, with some minor adjustments.

Transitional Period (1100s–1700s). In this era, tool making became a separate trade. People apprenticed themselves to the craft of making tools. Tools were capable of creating detailed carvings, textures, and complex shapes. Tools for creating joints and grooves for fitting pieces of wood were developed. Steel was used for the edges of most cutting tools.

Industrial Age (1800s). Here we witness the transitions from tools being created by a single craftsman to tools made by many in workshops and factories. The ability to cast and mold iron instead of shaping it by hand was developed. Standardization allowed parts to be interchanged and more tools were available to more people.

Modern Tool Period (1800s–Present). The development of electric power tools, automated machines, robots, and computers.

The American Industrial Hygiene Association (1996), in its guide to hand tools, indicates various categories of hand tools.

Wrenches. Used for tightening and loosening bolts.
Pliers. Used for bending, holding, or cutting.
Cutters. (Wire cutters) used for cutting.
Striking tools. Used for impacting, transferring force.

Struck or hammered tools. Used for driving, pounding, or swagging.
Screwdrivers. Used for turning screws.
Vises. Used for holding work pieces and materials.
Clamps. Used for compressing, holding in position, or binding together two or more parts.
Snips. Used for small cuts.
Saws. Used to cut or divide wood, bone, or metal.
Drills. Used to pierce, dig, or bore.
Knives. Used for incising materials.

The development of tools has allowed human cultures to develop and flourish. Humans have always depended on them to help accomplish many important tasks. Tools will play an important role in the future development of human culture and technology.

2 Human Factors and the Hand Tool

2.1 DISCUSSION

Pragmatism continually dominates human activities, particularly in those areas related to survival and accommodation to the physical environment. Hand tools have been an integral part of human cultures and human activities for multiple millennia. Hand tools must exist and will always exist. The important questions posed in this chapter are 1) why have hand tools been created, 2) why do they continue to exist, 3) how are they used by human beings?

2.2 HUMAN FACTORS AND ERGONOMICS

The rationale for the existence of hand tools within various human cultures is provided by the discipline known as human factors and ergonomics. This discipline studies the interface and the relationship of the human being and the various practical artifacts he/she has created for purposes of survival or purposes of leisurely enjoyment (Figure 2.1).

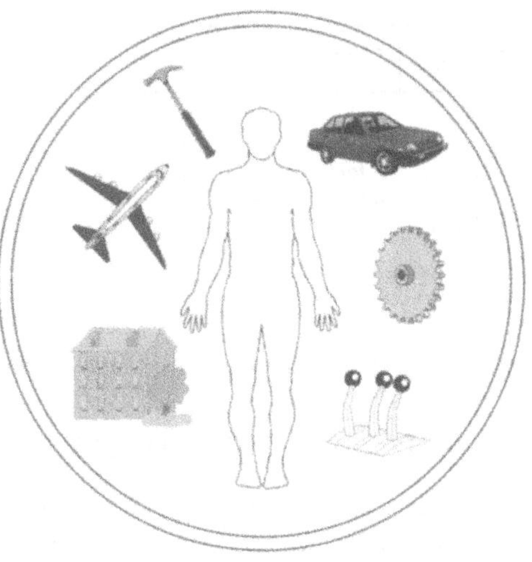

FIGURE 2.1 Human and Artifacts

These artifacts are physical and psychological extensions of the human being and compensate for many physical and psychological inadequacies which are attributed to the human being. It is important to attempt to provide a distinction between the terms ergonomics and human factors. Although there is a substantial overlap between the disciplines of human factors and ergonomics, and although practicing professionals in these fields deplore any attempts at distinguishing between these disciplines, a distinction will be offered for the sake of organizing this book. Human factors deals, though not exclusively, with psychological issues such as behavior, sensation, perception, information storage and retrieval, and decision making. Ergonomics deals, though not exclusively, with anatomical and physiological issues. This chapter will discuss the human factors aspects of the human/tool relationship. A future chapter will deal with the ergonomics aspect of the human/tool relationship (Figure 2.2).

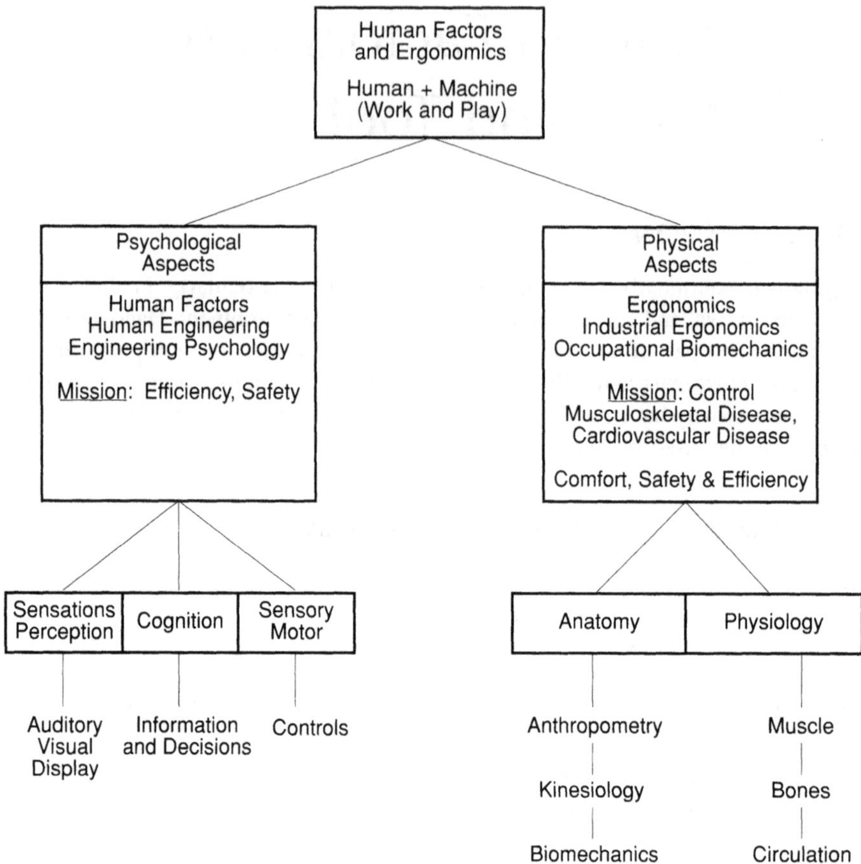

FIGURE 2.2 The relationship of the various subdisciplines of Human Factors and Ergonomics.

2.3 WHY HAVE HAND TOOLS BEEN CREATED?

Hand tools would not exist except for various practical human needs and requirements. Human factors authors and theoreticians such as Kantowitz and Sorkin (1983) and McCormick (1970), have spoken of the various human inadequacies and limitations, both physical and mental, which have led to the assignment of the relative responsibilities of the human and the "machine" in "human/machine systems." In essence, the human possesses several critical inadequacies (limitations), which must be compensated for by the machine. In general, the machine is superior to the human in such areas as strength, speed, vigilance, and endurance (Figures 2.3, 2.4, 2.5, 2.6).

Responsibilities and functions in various human/machine systems are maximized for the machine and minimized for the human with these inadequacies and limitations in mind. Hand tools, which may also be considered machines, compensate for human inadequacies and limitations while performing manual tasks. These insufficiencies are especially unique and are discussed in the following section. Some aspects of these discussions will be expanded in the chapter related to biomechanics.

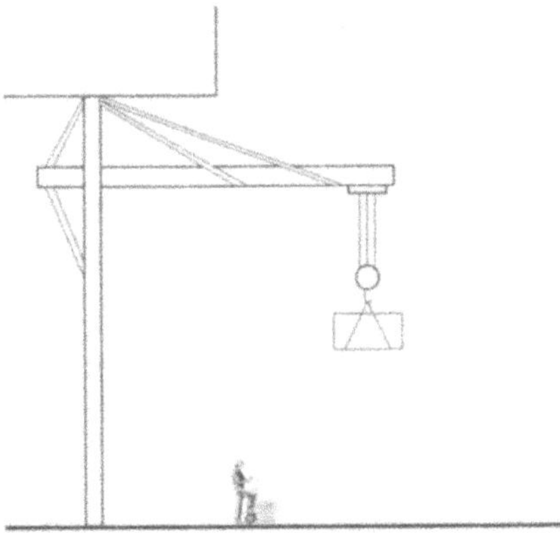

FIGURE 2.3 Crane. The artifact is strong.

FIGURE 2.4 Locomotive. The artifact is fast.

FIGURE 2.5 Production Machine. The artifact is tireless.

FIGURE 2.6 Control and Display Console. The artifact is vigilant.

2.3.1 Strength

Limitations in strength, particularly of the hand, have led to the design of hand tools which magnify human grip force (Figure 2.7).

FIGURE 2.7 Bolt Cutter. Tools magnify human strength. Pressure at handles A is greatly increased at B.

A 50th percentile male grip of approximately 100 pounds may be multiplied as much as three times by the handles of a pair of pliers or a pair of wire cutters. Two-handed tools such as pruning shears are capable of magnifying forces as much as ten times. The working ends of these tools provide a compressive force for purposes of crushing, holding, piercing, or cutting.

2.3.2 Penetrability

Human skin and underlying tissue are relatively soft and penetrable. Efforts in abrading most materials external to the human are generally unsuccessful without hand tools composed of some hard impenetrable material (Figure 2.8).

FIGURE 2.8 Wood Saw. The tool is harder and more impenetrable than human tissue.

These impenetrable materials are most traditionally iron and steel but may also be, or have been, wood, plastic or bone. Saws and files are the best examples of tools composed of hard materials which are capable of abrading softer materials.

2.3.3 Bluntness

Human appendages end in digits which have relatively blunt, broad ends. This bluntness has two disadvantages: 1) materials external to the human being cannot be pierced, cut, or penetrated by the fingers, 2) small objects cannot be readily grasped and manipulated by blunt fingers. Tools which compensate for this inadequacy are knives, scissors, chisels, awls, and drills which punch, drill, and cut various objects; and tweezers which grasp small objects (Figure 2.9).

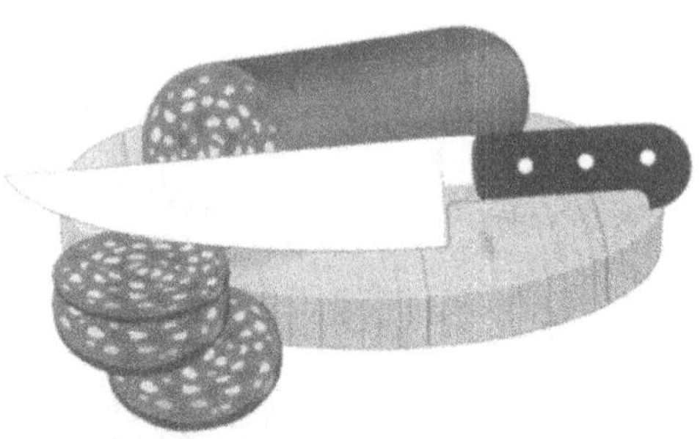

FIGURE 2.9 Carving Knife. The tool may be sharper and more pointed than human appendages.

2.3.4 Shortness

In some situations the human may wish to extend his/her reach towards a remote object or an object located in an inhospitable environment. Tools such as tongs and pruning poles extend human reach beyond the limits of arm hand length (Figure 2.10).

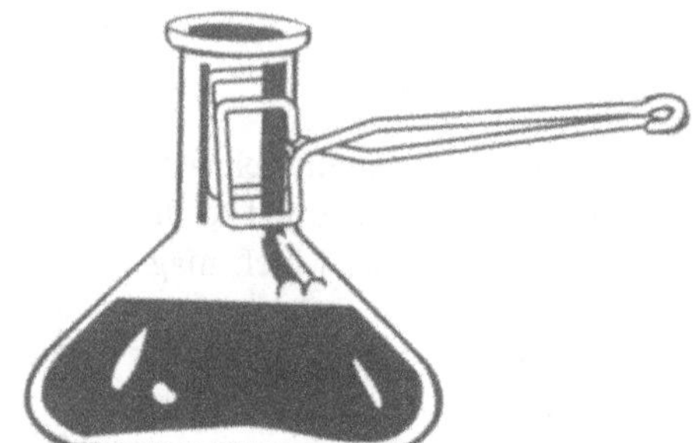

FIGURE 2.10 Tongs. Extension of human appendages. Tool reaching into inhospitable environment.

2.3.5 Flexibility

Hands and fingers may not retain rigidity due to fatigue and loss of strength over short periods of time. Tools, particularly tool handles such as those of a hammer, provide a needed rigidity which facilitates control and manipulation.

2.3.6 Limited Speed

Under some circumstances, the shortness of human limb limits the speed of manipulation of object and tools. Tools, which extend human reach, also may increase manipulation speed if manipulation movement is performed in an arcing motion (Figure 2.11).

The handle of a hammer allows a greater speed of motion of the hammer head as compared to holding the hammer head in the hand. The greater speed in turn produces a greater force during impact.

FIGURE 2.11 Hammer. The tool may increase speed of human appendages.

2.3.7 Summary

In summary, the essential hand tool characteristics which contribute to the human/tool relationship are mechanical force, hardness, extensibility, small surface areas and rigidity. These characteristics compensate for analogous human limitations and allow the human to perform a large number of manipulations of the immediate environment.

2.4 SYSTEMS

Human factors deals with the relationship of the human and any practical artifacts which the human may have created. This relationship, which always has a pragmatic goal, is referred to as a system. Systems must contain at least two components, one of which is a human being. Systems may be large and complex or small and simple (Figure 2.12).

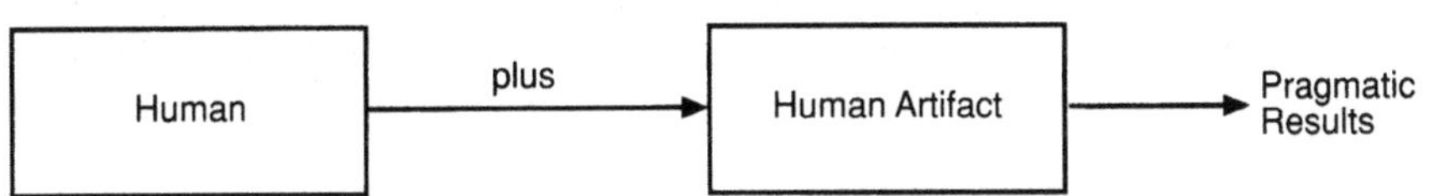

FIGURE 2.12 A system is composed of two or more components working together for a common goal. A human/artifact system contains at least one human being.

A human being using a hand tool may be considered to be a system.

2.4.1 Two Major Systems

Systems are categorized as Closed Loop or Open Loop Systems. A Closed Loop System is a continuous relationship in which the human constantly or intermittently, at regular intervals, directs (controls) the progress and status of a practical artifact. The artifact either maintains its status and condition or changes under the control of the human. The human receives feedback from the artifact and observes and monitors its progress to determine whether practical goals are being met. The human directs further changes in the system when necessary. There are many examples of Closed Loop Systems such as automobiles, aircraft, production machines and hand tools as well. In all cases the human has the opportunity for continuous direct or indirect control of the artifact and the artifact provides continuous direct or indirect feedback about its condition and status (Figure 2.13).

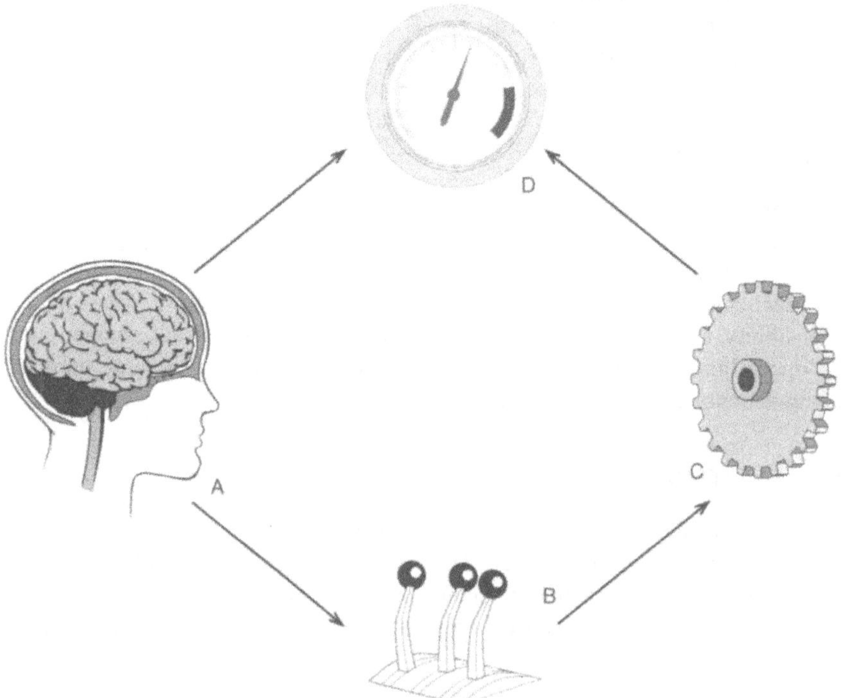

FIGURE 2.13 Closed Loop System. Human, A, monitors and changes system by activating Control B. Artifact, C, responds and communicates the change to human by way of Display D.

In contrast to the Closed Loop System, the Open Loop System is not a process manifesting continuous control by the human. After the artifact is deployed by the human, the artifact continues to function without any further control by the human. The most frequently cited example of an Open Loop System is the traditional artillery shell which, after discharge from the cannon, may be observed by the human but may no longer be controlled by the human.

2.4.1.1 Summary

A human being using a tool comprises a simple Closed Loop System in which the human provides continuous direct control of the tool and the tool provides continuous, elemental feedback to the human, by being observed by the human as the work is being performed.

2.4.2 Closed Loop Systems: Levels of Complexity

Kantowitz and Sorkin (1983) indicate that human/artifact systems can be classified into four levels of complexity. As the level of complexity progresses, the human is assigned fewer and fewer direct responsibilities within the system and the artifact is assigned more and more responsibilities. These levels are as follows:

2.4.2.1 Level 1

The human is in direct control and has direct physical contact with the artifact. The feedback from the artifact is simple and direct. Power and energy needed to run the system are provided by the human.

2.4.2.2 Level 2

The human is still in direct control and still in physical contact with the artifact. The feedback from the artifact is still simple and direct. Power and energy, however, are now provided by the artifact.

2.4.2.3 Level 3

The human is still in control, but the control over the artifact is indirect. This indirect control is usually manifested by knobs, push buttons, levers and other devices often referred to as controls. Feedback from the artifact is indirect and usually manifested by dials, gauges, lights and other devices often referred to as displays. The artifact still supplies the power and energy.

2.4.2.4 Level 4

The human has indirect minimal control after activating the system. There are still displays and controls, but the human is more a monitor, rather than operator, of the system. The artifact still supplies the power and energy.

2.4.3 The Human/Hand Tool System

In discussing these Closed Loop Systems from the complex to the simple, the fourth level may be referred to as an automated system. Examples of an automated system are automated manufacturing plants and auto pilots in aircraft. The third level is exemplified by complex, interconnected production sequences in factories, or various land, sea, and air vehicles where there is still substantial control by the human. The second and first levels are exemplified by hand tools (Figure 2.14).

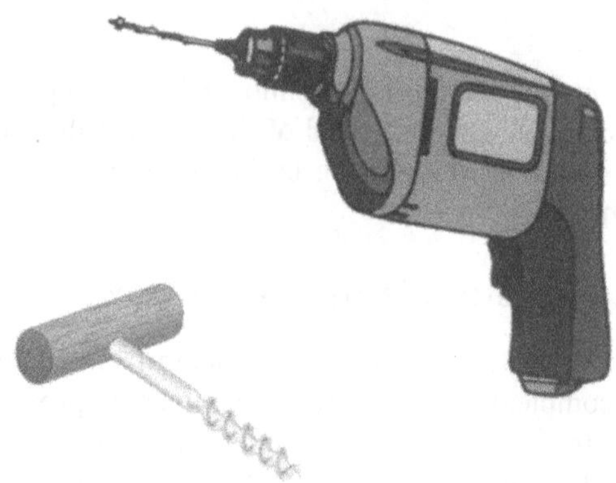

FIGURE 2.14 Drills. The cork screw below is human powered and an example of first level complexity. The drill above is electrically powered and an example of second level complexity.

In both second and first levels the human/tool relationship is physically direct and the human applies total control by placements of arm and hand. In both levels there is no direct feedback from the artifact in the form of dials, gauges, etc. Feedback to the human is usually direct in the form of a visual and tactile image describing the placement of the tool in relationship to the work piece. The first level which is the simplest of systems includes all basic unpowered tools such as knives, pliers, cutters, scrapers, drills, saws, etc. Energy and power are always supplied by the human. The second level includes any tool in the first level which is powered by electricity, compressed air, or any other energy form which is other than that supplied by the human. The energy usually supplies circular, rotating, reciprocating or longitudinal motion of the working part of the tool.

2.4.4 System Improvement and Evolution

Hand tools belong within the simplest, least complex and "primitive" categories of human/artifact systems. A question which arises is "can the need of such a primitive system be bypassed by more sophisticated, efficient, healthful systems?" The answer is negative. Barring the development of personal service robots, the need for direct human manipulation of the physical environment will always exist and humans, with their physical limitations, will always design and use hand tools to facilitate the manipulations. New tools may be conceived and created, methods of delivering power to powered hand tools may be improved, but the basic need for hand tools will never disappear.

2.5 SENSORY MOTOR CONSIDERATIONS

The human/hand tool relationship will always involve movement of hand, arm and tool. McCormick (1964) indicates five basic motor activities involved in work. These activities are applicable when the hand or hands are employing an unpowered or powered tool. The movements are as follows:

2.5.1 Positioning

Positioning is moving the hand and tool from one specific location to another. As an example: placing a screwdriver into the slot of a screw.

2.5.2 Repetition

Repeating many uniform movements of hand and tool. As an example: sawing wood, turning a screw driver.

2.5.3 Continuous

A long uninterrupted movement of the tool over the workpiece. As an example: making a long cut with a knife.

2.5.4 Serial

Movements involving several sequential uniform or non-uniform movements. As an example: striking a row of tacks with a hammer.

2.5.5 Static

An immobile frozen position of hand and tool while utilizing the tool. As an example: holding a workpiece with a pair of pliers.

2.6 SUMMARY

McCormick (1964) makes several statements about movements which can be applied to hand tool use.

All movements involving hand tools are performed under visual control which may sometimes be supplemented by tactile sensations.
Because of reaction time and overcoming of inertia, longer movements are relatively less time-consuming than shorter movements.
Inward and outward movements are faster than sideward movements.
Movements stopped by a mechanical barrier are faster than movements without such a barrier.

2.7 TRACKING

Hand movements with tools conform to the principles of tracking. Tracking is a continuous controlled movement of a tool towards a target. The target is the ultimate desired position of the tool. Individual movements of the tool towards the target are considered an input. The various positions of the tool on the way to the target are outputs. This form of tracking, when moving a tool, is referred to as pursuit tracking because the tool and the target, as well as their relative positions, are continually visible to the human. On the way to the target, the human will observe errors which

FIGURE 2.15 Tracking. Screwdriver moves towards screw, a series of deviations and compensations around a linear path.

are deviations from the correct path of the tool and will make corrections which compensate for these errors (Figure 2.15).

The corrections alternate from plus to minus positions and are mathematically defined as a zero order sinusoidal curve.

3 Epidemiology of Tool Use

3.1 INTRODUCTION

Epidemiology is the discipline which studies the distribution and the determinants of disease. Disease is defined as a state of departure from normal structures or functions within an organism. Epidemiology: 1) generally studies large groups of people, and 2) uses statistical techniques which provide quantified descriptions which, in turn, provide information, which may be used to control the incidence of disease. Epidemiology is usually identified with the medical profession; however, epidemiological approaches may be used by the safety and health professional who is often concerned with the incidence and distribution of injury and illness in the workplace or among members of the public. The following sections contain statistical information which relates to hand tools and other injury and illness issues as well. This information will indicate the relative importance or unimportance of the hand tool in relation to the overall efforts practiced by safety and health professionals.

3.2 BUREAU OF LABOR STATISTICS

Table 3.1 and Figures 3.1, 3.2 contain information extracted from 1996 data located in Tables 5 and 9, as provided by the Bureau of Labor Statistics. The most important facts follow:

1. Hand tools are associated with no more than 6.1% of all injuries and illnesses.
2. The areas of the anatomy generally considered most related to hand tools (wrist, hand, and fingers) are the site of no more than 17.1% of injuries and illnesses.
3. The areas of anatomy most related to hand tools (wrist, hand, and fingers) are related to 20 median days away from work which is 23% of all median days away from work (87) related to body parts. In most likelihood, a substantial portion of this 23% is not so much attributable to hand tools as it is to carpal tunnel syndrome and the use of keyboards.
4. Hand tools are associated with no more than four median days away from work which is 7% of all (55) median days away from work.

3.3 CONSUMER PRODUCT SAFETY COMMISSION

Table 3.2 and Figures 3.3, 3.4 contain information, provided by the CPSC's National Electronic Injury Surveillance System (NEISS), associated with injuries related to hand tools which were treated in hospital emergency rooms in 1996.

TABLE 3.1
Percent Distribution and Median Days away from Work for Non-Fatal Occupational Injuries and Illnesses by Selected Injury or Illness Characteristics. Private Industry, 1996.

Characteristic	Percentage	Median Days away from Work
Total: 1,800,000 Cases	**100**	**5**
	Nature of Injury or Illness	
Sprain, Strain	43.6	6
Bruise, Contusion	9.3	3
Cut, Laceration	7.1	3
Fracture	6.4	17
Carpal Tunnel Syndrome	1.6	25
Heat Burns	1.5	4
Tendinitis	.9	9
Chemical Burns	.6	2
Amputations	.5	20
Multiple Injuries	3.2	8
	Part of Body Affected	
Head	6.6	2
Eye	3.5	2
Neck	1.8	5
Trunk	38.1	6
Shoulder	5.1	8
Back	26.1	6
Upper Extremity	22.6	6
Wrist	5.0	12
Hand, No Finger	4.0	4
Finger	8.1	4
Lower Extremity	20.0	6
Knee	6.8	9
Foot, No Toe	3.6	5
Toe	1.2	4
Body Systems	1.4	3
Multiple Parts	8.6	7
	Source of Injury	
Chemicals	1.9	2
Containers	14.5	6
Furniture, Fixtures	3.6	5
Machinery	6.6	6
Parts and Materials	11.1	5
Worker Motion, Position	14.5	8
Floors, Walkways	16.2	7
Tools, Instruments, Equipment	6.1	4
Vehicles	8.0	7
Health Care Patients	4.6	5

TABLE 3.1 (continued)
Percent Distribution and Median Days away from Work for Non-Fatal Occupational Injuries and Illnesses by Selected Injury or Illness Characteristics. Private Industry, 1996.

Characteristic	Percentage	Median Days away from Work
Total: 1,800,000 Cases	**100**	**5**
	Event or Exposure	
Contact with objects	26.2	4
Struck by object	12.7	4
Struck against object	6.8	4
Caught in equipment	4.2	7
Fall to lower level	5.2	10
Fall same level	11.7	7
Slips, Trips, No Fall	3.2	6
Overexertion	28.0	6
Overexertion lifting	16.6	6
Repetitive Motion	3.9	17
Harmful Substance	4.6	3
Transportation Accident	4.1	9
Fire, Explosion	.2	8
Assault, Violence	1.0	5

The above data extracted from Tables 5 and 9 of the U.S. Bureau of Labor Statistics Website.

National estimates are provided based upon sampling procedures. The hand tools related to injury are ranked according to frequency. The majority of those visiting the emergency room were consumers rather than workers. The following facts are provided:

1. Knife, hand saw, hammer, manual scissors, and chain saw are those tools most often related to an emergency room visit.
2. The age classification experiencing the greatest incidence is ages 25 to 64.
3. The vast majority of visits ended in being released rather than being transferred or hospitalized.
4. A total of 313,600 visits were counted of which 17,300 were related to various tools, thus tools accounted for 5.5% of all visits.

3.4 ADDITIONAL STUDY

Mital and Kilbom (1992) report a study, by Aghazada and Mital (1987), based upon a questionnaire related to workers and hand tools submitted to governmental agencies. Some pertinent facts are provided:

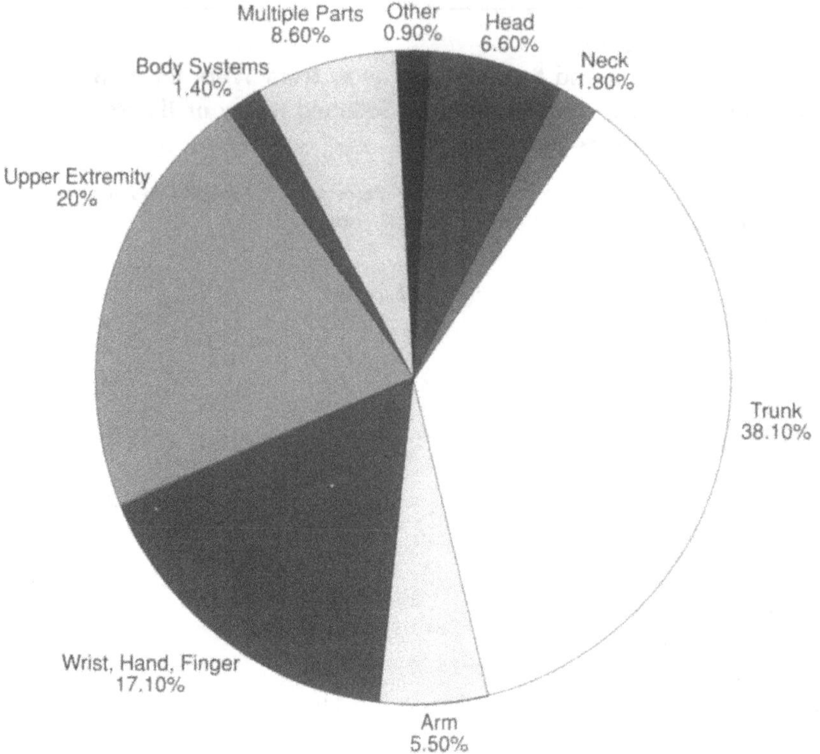

FIGURE 3.1 Percentage of Injury Frequency by Body Part, Year 1996, Extracted from U.S. Bureau of Labor Statistics.

1. Hand tool injuries account for 9% of industrial injuries.
2. Non-powered hand tools were related to 3.9% of amputations and powered tools 5.1% of amputations.
3. Knives produced the most severe injuries among non-powered tools and saws, the most severe injuries among powered tools.
4. 80% of tool injuries are caused by non-powered rather than powered tools.
5. Non-powered tools producing the greatest number of injuries are knife, hammer, wrench and shovel.
6. Powered tools producing the greatest number of injuries are saws, drills, hammers, and grinders.
7. Injuries related to non-power and powered tools are mostly caused by striking against or being struck by the tool.
8. Injuries resulting from all tools are predominantly cuts, lacerations, sprains and strains.
9. 30% of injuries related to tools occur to the fingers.
10. The greatest proportion of injuries occur to the 18 to 35 year age group.

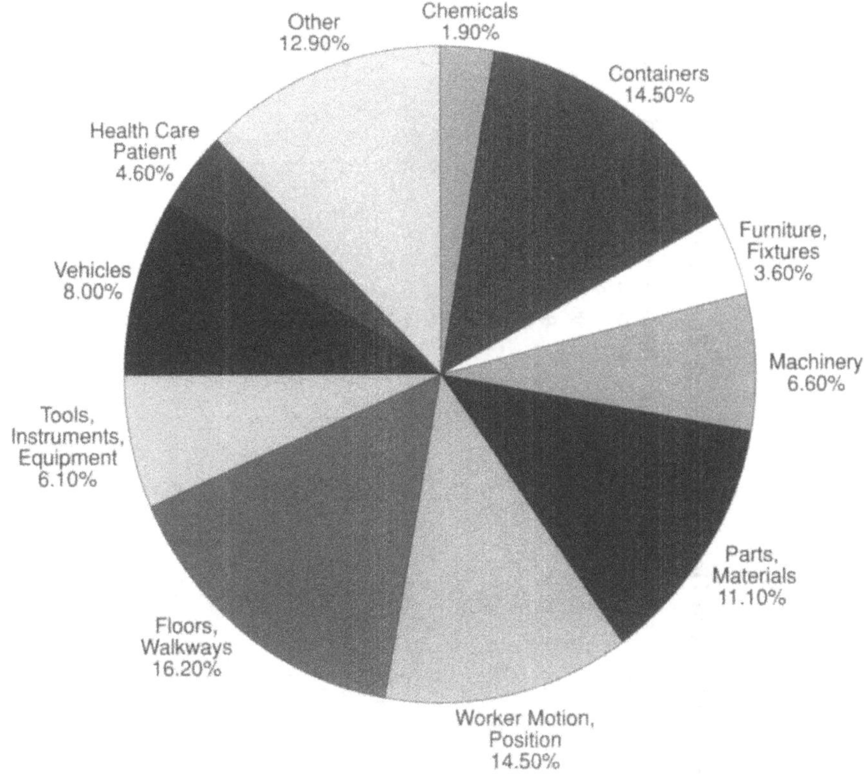

FIGURE 3.2 Percentage of Frequency of Injury by Injury Source, Year 1996. Extracted from U.S. Bureau of Labor Statistics.

3.5 GENERAL CONCLUSIONS

Some broad conclusions may be derived from the prior information.

1. Percentage-wise, hand tools are not a relatively frequent source of trauma or illness among workers or consumers.
2. Non-powered hand tools are relatively a more frequent source of trauma than powered tools.
3. The non-powered knife, hammer, saw, scissors, and wrench are the most frequent contributions to trauma.
4. Injuries to hand tools and workers and consumers are concentrated in the 25 to 35 year age group.

Although hand tools are not as critical an area of safety and health as other areas, such as manual materials handling VDT operations, slips, trips, and falls, measures should be taken in all safety programs to control trauma and disease from this source.

TABLE 3.2
National Electronic Injury Surveillance System Product Associated Visits to Hospital Emergency Rooms, 1996

Product	National Estimate	Coefficient of Variation	Sample Count	Age 0–4%	Age 5–14%	Age 15–24%	Age 25–64%	Age 65+%	Treated and Released %	Transferred %	Hospitalized %
0464 Knife	444,604	.07	10,711	3.1	12.9	21.3	56.8	5.8	99.2	.2	.4
0830 Hand Saws	89,786	.08	1,988	.4	3.8	10.9	66.4	18.3	94.5	1.8	3.8
0827 Hammers	44,828	.08	972	2.3	11.0	14.3	68.7	3.7	99.3	.3	.2
0420 Manual Scissors	30,342	.08	772	11.4	32.4	13.9	36.4	5.8	98.7	.4	.8
1411 Chain Saw	35,132	.09	657	0	1.7	17.1	70.8	10.4	96.5	.5	3.0
0847 Drills	16,847	.10	360	1.5	1.0	13.2	73.7	10.6	99.8	0	.2
0857 Pliers, Wire Cutters, Wrenches	14,543	.10	317	2.0	5.2	15.2	72.9	4.8	99.9	0	.1
1426 Hatchets, Axes	14,332	.12	275	1.2	14.8	16.2	64.3	3.6	98.7	0	1.3
0828 Screw Drivers	11,629	.11	270	8.8	8.7	15.3	60.0	7.2	99.7	.3	0
0832 Portable Circular Saw	11,115	.12	257	0	3.1	13.9	65.9	17.1	89.2	3.2	7.6
0865 Portable Power Grinders, Buffers	14,443	.11	247	.1	1.7	15.6	78.1	4.5	98.9	0	1.1
0882 Nail Guns, Stud Drivers	9,013	.15	176	0	1.2	18.7	78.6	1.4	93.5	2.5	4.0
0870 Awl, Chisel, Plane	8,222	.13	167	.5	3.2	12.6	70.5	13.1	97.7	0	2.3
0862 Manual Filing, Sanding Tool	2,311	.22	51	0	0	23.6	65.1	11.2	98.4	0	.7
0834 Heavy Duty Stapler	1,704	.18	50	6.0	15.0	15.4	61.5	2.0	100	0	0

Information provided by the Consumer Product Safety Commission's National Electronic Injury Surveillance System.

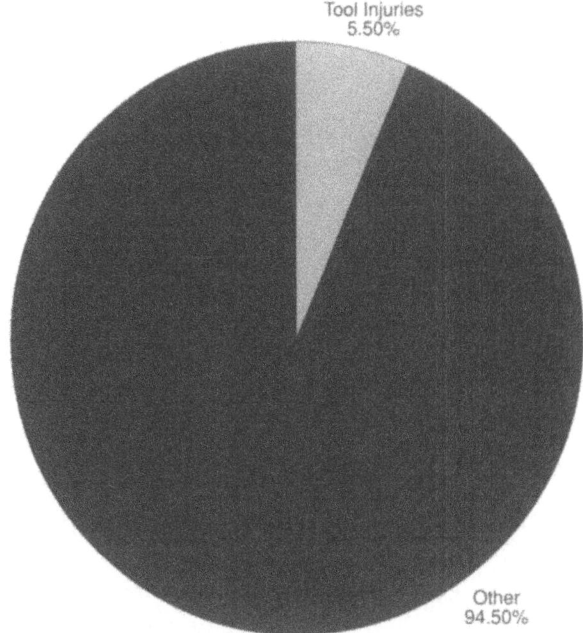

FIGURE 3.3 Percentage of Emergency Room visits in 1996 Related to Injurious Objects. Extracted from U.S. Consumer Product Safety Commission.

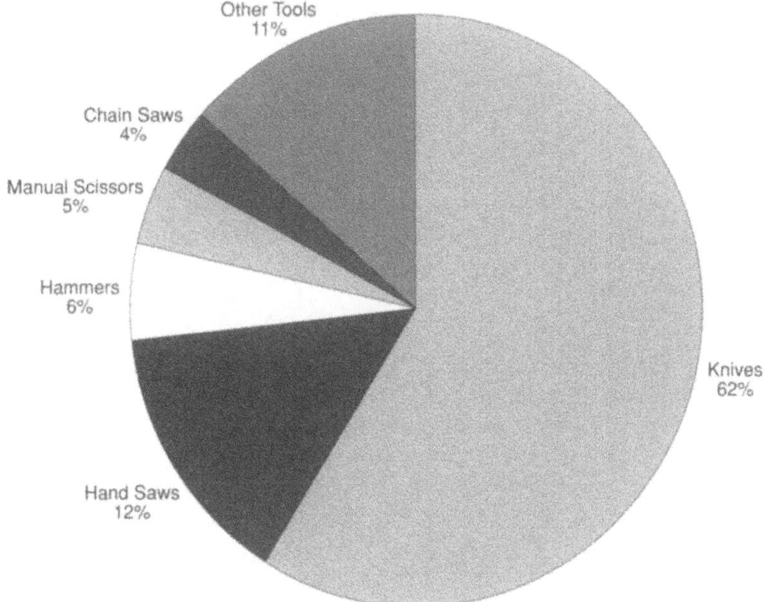

FIGURE 3.4 Percentage of Emergency Room visits in 1996 Related to Injury by Tool. Extracted from U.S. Consumer Product Safety Commission.

4 Pathology and Physiology of Hand and Arm

4.1 DISCUSSION

As indicated in other sections of this book, poor hand tool design may lead to two categories of trauma: instant (acute) trauma or cumulative (subacute) trauma. Instant trauma occurs immediately when a hand tool intrudes upon human tissue. Causation of instant trauma is discussed in the chapter related to hand tool safety. Cumulative trauma develops gradually when a tool is used over a period of time. The immediate contributing causes of cumulative trauma are discussed in the chapter related to ergonomics. The pathological results of these two forms of trauma are discussed in this chapter.

4.2 MUSCULOSKELETAL DISORDERS

Musculoskeletal disorders (diseases) involve muscle, bone, cartilage, tendon, and ligament, and affect the locomotion (movement) of human beings. Musculoskeletal disorders may also affect parts of the nervous system locally related to the musculoskeletal disorder. Various disorders will be described below.

4.2.1 Types of Musculoskeletal Disorders

Parker and Imbus (1992) and Naderi and Ayoub (1991) provide information about various disorders related to hand and arm. In general these disorders are best categorized as 1) tendon disorders, 2) nerve disorders, and 3) neurovascular disorders.

4.2.1.1 Tendon Disorders

These disorders generally occur in the area of the tendon where the tendon inserts into a bone and proximally to a joint. The disorders are associated with local aches or pains, tenderness and possible swelling due to edema (fluid buildup).

Tendinitis. This tendon disorder is an inflammation and the result of substantial tension placed upon the tendon beyond its normal tensile strength. Some fibers which comprise the tendon may fray away from the main body. Calcification may occur in those tendons which have no sheaths. If not enough rest is provided the tendon may become permanently weakened.

Tenosynovitis. Tenosynovitis occurs in those areas where a tendon is covered by a sheath (synovium). Under conditions of extreme repetition, excessive synovial fluid is produced. An accumulation of synovial fluid leads to pain and swelling.

Stenosing Tenosynovitis. This disorder is characterized by a constriction of the tendon sheath in the area of the inflammation. DeQuervain's Disease is the most prominent of these disorders. Caused by repetition, it occurs at the base of the thumb to the tendons of the thumb which connect to the forearm. An additional prominent form of stenosing tenosynovitis is stenosing tenosynovitis crepitans (Trigger Finger). Swelling of a tendon, due to mechanical pressure from a sharp hand tool handle, causes a locking of the tendon within the sheath, resulting in jerking movements of the finger.

Ganglionic Cyst. This disorder of tendon sheath is manifested by a cyst (sack filled with fluid) protruding through the skin at the dorsal aspect of the wrist.

Epicondylitis. This condition is a tendinitis of unsheathed tendons emanating from the finger-extending muscles of the forearm that insert into the humerus. The condition occurs from excessive tensile force. Epicondylitis of the lateral aspect of the elbow is referred to as Tennis Elbow. Epicondylitis at the medial aspect of the elbow is referred to as Golfer's Elbow.

Rotator Cuff Tendinitis. This disorder is a tendinitis caused by stresses involved with the four tendons which rotate the arm at the shoulder.

4.2.1.2 Nerve Disorders

Damage or pressure upon nerves may occur where there is a proximity of a nerve to a swollen inflamed tendon. Frequently, the pressure will cause symptoms in areas distal to the area where the injury has occurred.

Carpal Tunnel Syndrome. Excessive tensions upon the finger flexing tendons which pass through the carpal tunnel will cause tenosynovitis of the tendons with a resultant swelling. The swelling in turn places pressure upon the median nerve which also passes through the carpal tunnel. Pressure upon the median nerve causes tingling, numbness, and pain in the first three fingers and base of the thumb.

4.2.1.3 Neurovascular Disorders

These disorders involve nerves and blood vessels as well. Circulatory impairment in addition to neuropathic conditions occur.

Thoracic Outlet Syndrome. This disorder will display symptoms similar to carpal tunnel syndrome. Continuous muscular tension, during abduction of the arm, causes compression upon the neurovascular bundle passing through the outlet between the thorax and upper arm.

4.2.2 General Etiology: Cumulative Trauma

Cumulative trauma is a gradual, insidious buildup of trauma, characterized by many biomechanical insults to tissues over a period of time. Armstrong (1983) and Putz-Anderson (1988) have indicated a general causality related to cumulative traumas that result in musculoskeletal disease. This causality, which is usually accepted by the safety

FIGURE 4.1 Cumulative trauma variables. Force, repetition, posture, and duration.

and health profession, involves four basic interacting variables: (Figure 4.1) posture, force, repetition, and duration. These variables may be related to hand tools.

4.2.2.1 Posture

The positional interface (posture) of human hand with tool, as well as human arm with hand and tool, will determine whether maximal or minimal biomechanical stresses are being placed upon the anatomy. Excessive biomechanical stress, in turn, will lead to cumulative trauma. Biomechanical stresses related to the hand tool cannot be totally eliminated but may be at least controlled and kept to a minimum. An application of the principles of leverage, as described in the chapters on ergonomics and biomechanics, will provide optimal postures which the hand should assume when using the tool. Less than optimal postures lead to unnecessary biomechanical stress.

4.2.2.2 Force

Excessive stresses occur when less than optimal postures related to the tool are assumed. These excessive stresses may be further compounded when large forces are applied during movements, or heavy loads are supported while holding an immobile position.

4.2.2.3 Repetition

Insults to tissues occur when biomechanically disadvantageous postures are magnified by excessive forces. This circumstance, though perilous, may not cause injury if these positions and forces are infrequently assumed. Tissues possess great propensity for rejuvenation, provided adequate resting periods are provided. If, however, those positions and forces are repeated constantly over short periods of time, there is no opportunity for rejuvenation and temporary impairment may occur.

4.2.2.4 Duration

A forceful biomechanically disadvantageous position repeated frequently for a short period of time may lead to a temporary impairment. A temporary impairment may be alleviated by eventual rest. If, however, stressful repetitions occur over a prolonged duration of time such as weeks, months, or years, then a serious permanent disorder will occur.

4.2.2.5 Interrelationship

The four aforementioned variables are interactive and, in most circumstances, musculoskeletal disorders will not occur unless all four variables are present during hand tool use. A question arises related to the relative importance of the four variables. Logically, the most important, critical variable is posture. The other three variables have no meaningful effect if a disadvantageous posture is not assumed. This opinion is further supported by McCauley Bell and Crumpton (1997) who, using a panel of experts, developed a fuzzy linguistic model for predicting carpal tunnel syndrome. These authors reported relative weights for task-related factors as awkward joint posture .327, repetition .206, hand tool use .196, force .136, and task duration .135. This would lead to a conclusion that posture and repetition are considered to be a primary etiology in cumulative trauma.

4.3 INSTANT TRAUMA

Hand tool use may cause various instantaneous injuries to hand and wrist. Some of the injuries are described below.

Fractures. Bones may be broken as the result of mishaps related to the use of hand tools. Fractures are often simple fractures but may also be compound fractures in which ends of the broken bone protrude through the skin. Other forms of fractures are comminuted (shattered), impacted (one end of bone driven into the other), and greenstick (incomplete, splintered).

Dislocations. Dislocations are complete displacements or misalignments of the opposing surfaces of a joint. Dislocations may be total or partial. Dislocations frequently occur to the fingers because of their relative fragility and their involvement with hand tools.

Sprains. Spraining is a temporary displacement or misalignment of a joint which causes tearing of the ligaments of the joint. Fingers are vulnerable to sprains as well as the above mentioned dislocations.

Strains. Strains involve stretching or tearing of muscle fibers. Tendons may also become involved with strains. The condition may, on rare occasion, involve a severe muscle rupture.

4.4 VIBRATIONS

A vibration is a rapid, repeated, oscillating movement of an elastic material departing from and returning to its resting position. Vibrations may occur in gaseous, liquid, or solid media, but may not occur in a vacuum. The number of oscillations which occur in one second is referred to as frequency. Frequency is usually stated as hertz (H_z). The amount of movement (displacement) away from the resting position, as well as its velocity and acceleration, will describe the magnitude of the vibration. This magnitude is often stated in gravity (g) or in m/s^2. A primary vibrating object (induction force), in contact with another solid elastic object, may cause the second object to vibrate (forced vibration). A substantial amount of forced vibration (resonance) of the second object may occur only if the frequency of the primary vibrating object approximates the natural frequency of the second object. The natural frequency of an object is dependent upon its configuration and mass.

4.4.1 Areas of Concern

The major concerns related to vibrations are 1) whole body vibrations, and 2) segmental vibrations. Whole body vibration is generated through contact of buttocks and feet with such vibration objects as vehicles and machines. These vibrations may affect the circulatory, urinary, and central nervous system, and display symptoms such as fatigue, insomnia, and headache.

The major concern of segmental vibrations is in the hands. Hands contacting vibrating power tools, such as chain saws, chipping hammers, and grinders, for extended periods of time are vulnerable to Raynaud's Disease (white fingers) or other cumulative traumas to wrist or elbow. Raynaud's Disease is typified by damage to blood vessels and nerves within the tissues of the hand. Symptoms include blanching of fingers, numbness, and difficulty in their use, particularly in cold surroundings. Eastman Kodak (1983) indicates that the hand may be affected by vibration magnitudes of 1.5 g to 80 g and frequencies of 8 to 500 H_z. Raynaud's is particularly associated with 1.5 g to 80 g and 25 to 150 H_z.

4.4.2 Vibration Control

Disease from vibrating powered hand tools is not a frequent occurrence. In the event of uncertainty regarding particular hand tools, consulting firms specializing in vibration testing may be hired. These firms may pinpoint offending tools and provide allowable time exposures for the use of a particular tool. Vibration hazards are usually controlled by the following methods:

1. Controlling time exposures to the tool in question. The less the tool is handled the less the probability of disease.
2. Using damping designs, such as springs or enclosed gas, within the operating parts of the tool.
3. Using damping resilient materials in the handle of the tool.
4. Using damping personal protective equipment such as gloves.
5. Using robots or remote controls to obviate the need of handling vibrating tools.

5 Anatomy of the Upper Extremity

5.1 OVERVIEW

An understanding of the physical interface of the human and the tool requires a knowledge of the anatomy of those parts of the human which touch the tool as well as those parts of the human which are proximal to the tool. Anatomy is the discipline that studies the structure and configuration of organisms, such as the species *Homo sapiens*. Anatomy is divided into the following major subdisciplines.

Gross Anatomy. The study of an organism's structure by use of the naked eye.

Microscopic Anatomy. The study of an organism's structure by use of magnification devices.

Comparative Anatomy. The comparison of analogous anatomical structures between species or within a species.

Surface Anatomy. The study of the external contours and landmarks of an organism.

Systematic Anatomy. The methodical, system by system, study of an organism's structures. (A system is defined as a group of components, such as organs, working together to perform a particular function.)

Regional Anatomy. The methodical, region by region, study of an organism's structures. (A region is defined as a readily recognizable portion of an organism.)

Questions of gross, microscopic and comparative anatomy are minimally addressed in this book. This chapter will initially deal with regional, surface anatomy and, in more depth, with systematic anatomy.

5.1.1 Regional and Surface Anatomy

The major regions of the human anatomy are head, neck, trunk, lower extremities (legs) and upper extremities (arms). The left or right upper extremity is divided into subregions.

These subregions are as follows:

Shoulder (pectoral girdle).
Upper Arm (brachium).
Forearm (antebrachium).
Wrist (carpus).
Hand (metacarpus or manus).
Fingers (phalanges).

There are boney protrusions located upon the upper extremity. These boney protrusions are often detectable by visual inspection and always detectable by palpation (examination by using finger pressure) (Figure 5.1). These landmarks are as follows:

Acromion. An elevated protrusion at the top of the shoulder.
Olecranon. Protrusion at the elbow joint.
Radial Styloid. Protrusion at the top of the forearm below the thumb.
Ulnar Styloid. Protrusion at the top of the forearm below the smallest finger.
Metacarpals (knuckles). Numbered one to five starting at the thumb.
Dactylions (finger tips). Numbered one to five starting at the thumb.

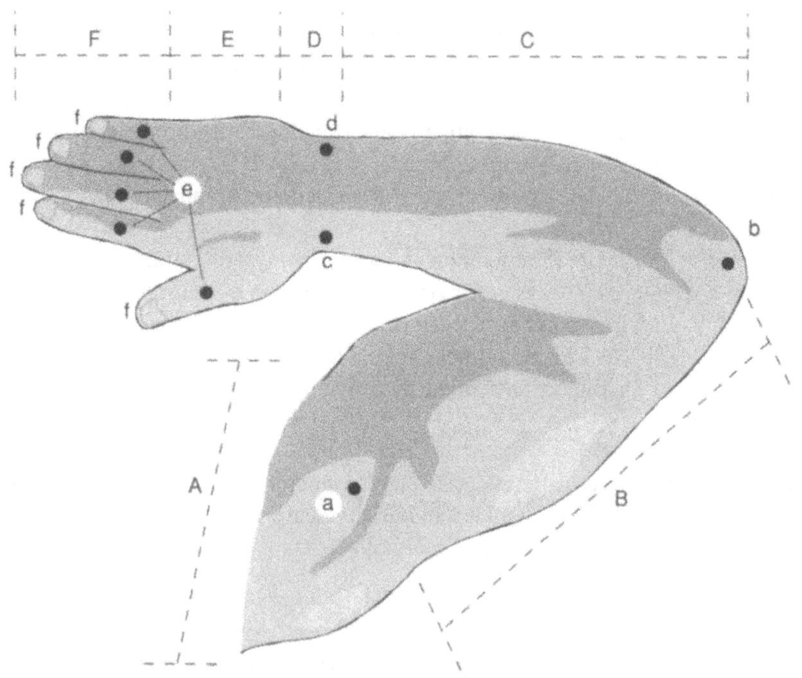

FIGURE 5.1 Upper Extremity Posterior View

Anatomical Landmarks:

A. Shoulder	a. Acromion
B. Upper Arm (brachium)	b. Olecranon
C. Forearm (antebrachium)	c. Radial Styloid
D. Wrist (carpus)	d. Ulnar Styloid
E. Hand (metacarpus, manus)	e. Knuckles (metacarpals)
F. Fingers (phalanges)	f. Finer Tips (dactylions)

5.1.2 Reference Points

Anatomists have devised a number of reference planes and directional views, which are used as guides for locating and describing human anatomical structures (Figure 5.2).

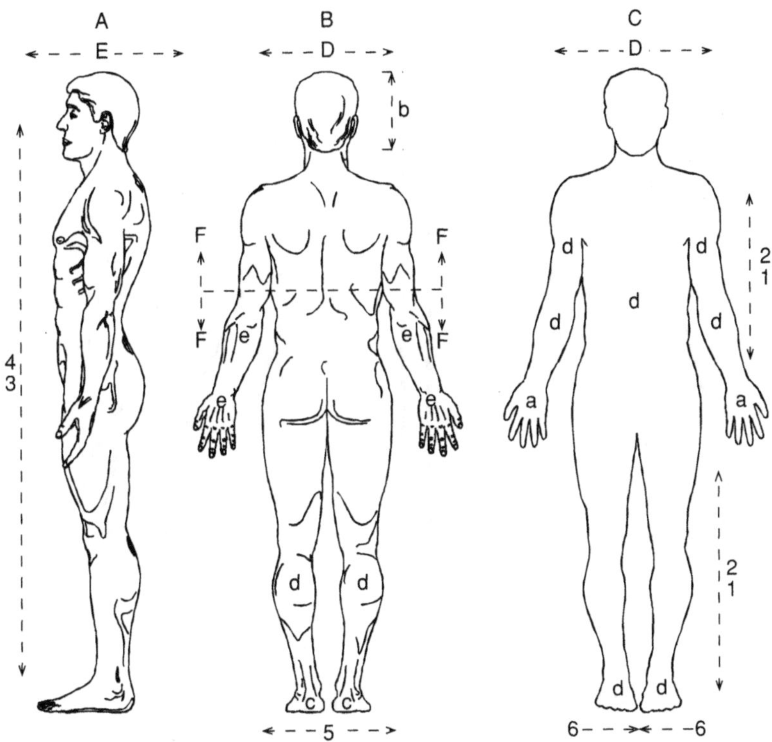

FIGURE 5.2 Reference Points, Cadaver Position

A. Lateral View
B. Posterior View
C. Anterior View
D. Coronal Plane
E. Sagittal Plane
F. Transverse Plane

1. Distal
2. Proximal
3. Inferior
4. Superior
5. Lateral
6. Medial

a. Palmar
b. Cranial
c. Plantar (foot, sole)
d. Ventral
e. Dorsal

These guides are based upon an initial posture of the human anatomy commonly known as the cadaver position. In the cadaver position the human is usually portrayed standing upright with feet together pointing straight forward and arms at the side with palms facing forward. The following are the most commonly used reference terms which generally apply to the entire anatomy. These terms may also be specifically applied to the upper extremity (Figure 5.3).

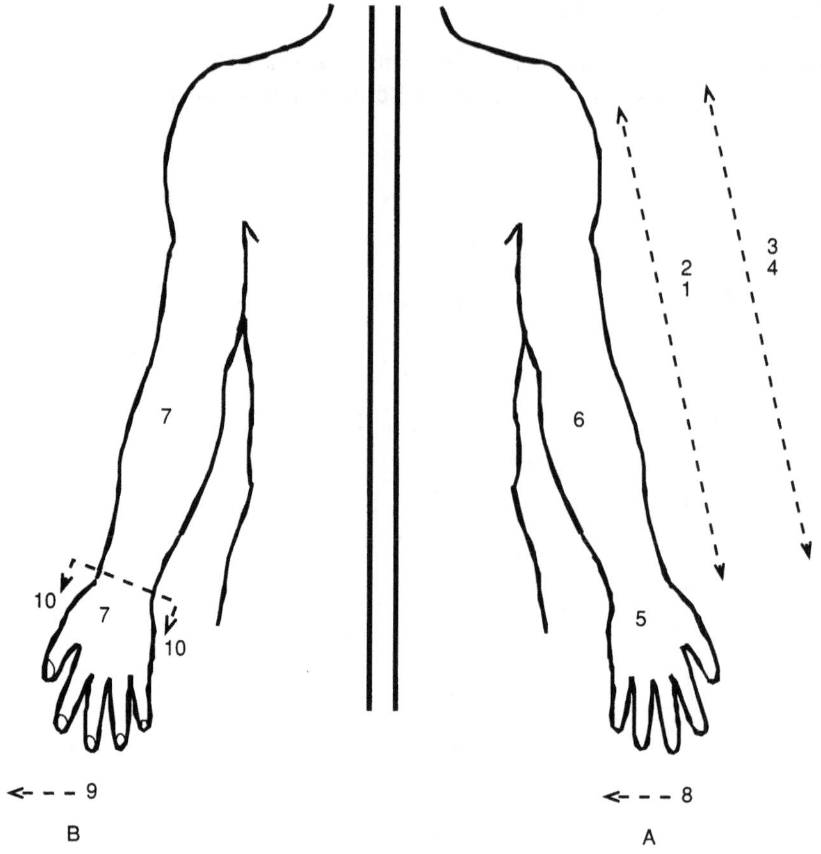

FIGURE 5.3 Reference Points, Upper Extremity

A. Left Arm/Anterior View
B. Left Arm/Posterior View
1. Distal
2. Proximal
3. Superior
4. Inferior
5. Palmar
6. Ventral
7. Dorsal
8. Medial
9. Lateral
10. Transverse Plane

Distal. Farther or farthest, especially from a point of attachment.
Proximal. Nearer or nearest, especially to a point of attachment.
Superior. Above, especially towards the head.
Inferior. Below, especially away from the head.
Palmar. A view towards the palm of the hand.
Ventral. A view of the softer aspects (belly) of arm and forearm.
Dorsal. A view of harder aspects (back) of hand, arm, and forearm.
Anterior. A view at the front of a structure.
Posterior. A view at the rear of a structure.

Medial. Towards the longitudinal midline of the trunk.
Lateral. Away from the longitudinal midline of the trunk.
Transverse Plane. A horizontal cross-sectional view of the trunk or an appendage.
Sagittal Plane. A front/back longitudinal cross-sectional view of the trunk or an appendage.
Coronal Plane. A side-to-side longitudinal cross-sectional view of the trunk or appendages.

Of the above terminologies, it should be noted that the sagittal and coronal planes do not, very frequently, apply to the upper extremity. It should also be noted that there is an interchangeability of terms such as distal and inferior, proximal and superior, palmar, ventral, and anterior, and dorsal and posterior. The following are some statements illustrating the use of various terminologies while the human anatomy is in the standard cadaver position.

The fifth finger is medial to the thumb.
The thumb is lateral to the fifth finger.
The hand is distal to the elbow and inferior to the elbow.
The elbow is proximal to the hand and superior to the hand.
When observing the palm of the hand, the hand provides a palmar view, as well as, an anterior view.
When observing the palm of the hand, the arm and forearm provide a ventral, as well as, anterior view.
When observing the back of the hand, the hand, arm, and forearm provide a dorsal, as well as a posterior view.
A CAT (Computerized Axial Tomography) scan image of arm or forearm provides a transverse view of arm or forearm.

5.2 SKELETAL SYSTEM

The human anatomy contains 200 skeletal bones and additionally six small bones located within the middle ears. The major segments of the human skeleton are the axial skeleton which comprises the central part of the anatomy and the appendicular skeleton which comprises the attached arms and legs (Figure 5.4).

The regions are further subdivided into components and subcomponents. Table 5.1 contains regions, components and subcomponents of the skeleton, and provides the number of bones each contains.

5.2.1 FUNCTIONS

There are several unique and essential functions which are attributed to the skeleton. These functions may be classified as physical or biochemical in nature.

Physical functions include structural support of the entire anatomy or parts of the anatomy. This structural support provides a rigid framework which prevents the collapse of surrounding soft tissue. In this regard, the skeleton's legs, spine, and neck allow a human to maintain an erect posture. The skeleton also physically provides protective barriers which cover vulnerable soft tissues. Examples of protective skeletal barriers are the cranium, which encases the brain, and the ribs, which encase the lungs and heart. Finally, the skeleton's physical functions include the utilization of bones as levers which give rise to fine movements, such as manipulation, and grosser movements, such as walking.

Biochemical functions include hemopoiesis which is the production within the marrow of red and white blood cells and blood platelets. Biochemical functions also include the storage of some essential minerals, such as calcium and phosphorous, which may be used systemically at time of need.

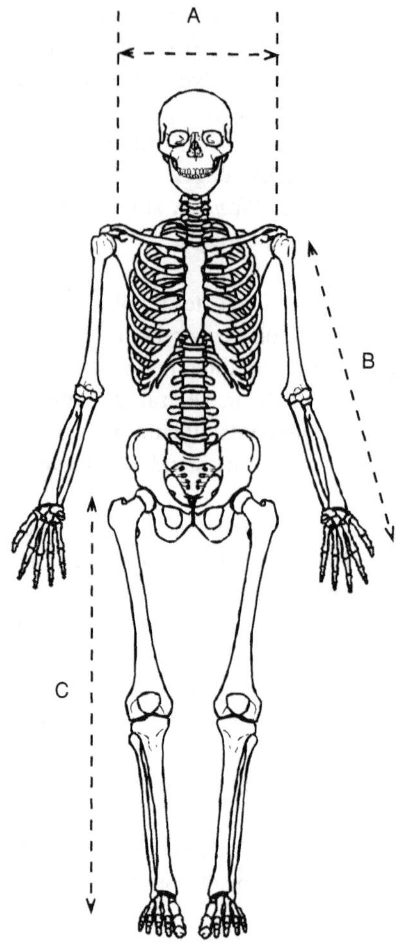

FIGURE 5.4 Skeleton, Anterior View

A. Axial Skeleton: Skull, Spine, Ribs
B. Appendicular Skeleton: Arms
C. Appendicular Skeleton: Legs

5.2.2 Types of Bones

There are four major categories of bones within the human skeleton: long bones, short bones, flat bones, and irregular bones.

Long bones are substantially longer in one dimension than another, are often associated with leverage and motion, and are often discovered in the upper and lower extremities.

Short bones have two dimensions which are approximately equal, are approximately square or cubic in shape, and transfer forces from other bones.

Flat bones have thin cross sections and are long and wide. These bones are used as protective shields, or as anchoring areas into which muscle tissue is attached.

Irregular bones have many faceted and unusual shapes and a variety of functions.

TABLE 5.1
Human Skeletal System

Segment	Component	Sub-component	Number of Bones
Axial Skeleton	Skull	Cranium	8
		Face	14
		Hyoid	1
	Trunk	Vertebrae	26
		Ribs	24
		Sternum	1
Appendicular Skeleton	Lower Extremity (legs)		62
	Upper Extremity (arms)		64
			200

5.2.3 Joints

Invariably, the endings of bones meet the endings of other bones at areas known as joints or articulations. There are three major varieties of joints within the skeletal system, which are the synarthrodic, amphiarthrodic and diarthrodic joints.

Synarthrodic joints are rigid, immovable close-fitting junctions (sutures) between bones. Examples of such joints are those found in the skull where the component bones of the cranium join together to form the entire cranium.

Amphiarthrodic joints are slightly moveable spaces between bones and separate the bones with a flexible fibrous or cartilaginous material. Examples of these joints are found in the spine. The spaces between spinal vertebrae are filled in by intervertebral discs thus forming an amphiarthrodic joint.

Diarthrodic (synovial) joints are the most commonly occurring joints of the skeletal system and are the only variety of joints found in the upper extremity. Diarthrodic joints are flexible and provide a wide range of motion. A typical diarthrodic joint is a cavity which separates the cartilaginous ending of two bones. The cavity is filled with a lubricating synovial fluid contained within a fibrous capsule. The entire joint is held in correct alignment with fibrous bands known as ligaments.

5.2.4 Bones in the Upper Extremity

These are considered bones of the appendicular skeleton and include the shoulder, upper arm, forearm, wrist, and hand (Figure 5.5 and Table 5.2).

TABLE 5.2
Bones of The Upper Extremity

Bone	Type	Number
Scapula	Flat	1
Clavicle	Long	1
Humerus	Long	1
Ulna	Long	1
Radius	Long	1
Carpal: scaphoid	Short	1
Carpal: lunate	Short	1
Carpal: triquetrum	Short	1
Carpal: pisiform	Short	1
Carpal: capitate	Short	1
Carpal: hamate	Short	1
Carpal: trapezium	Short	1
Carpal: trapezoid	Short	1
Metacarpals	Long	5
Phalanges	Long	14
		32

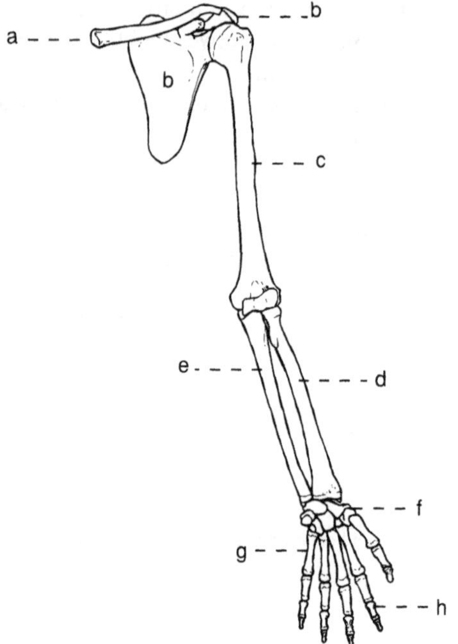

FIGURE 5.5 Skeleton, Upper Extremity

Anterior View:

a. clavicle
b. scapula
c. humerus
d. radius
e. ulna
f. carpals
g. metacarpals
h. phalanges

5.2.4.1 The Shoulder

The bones located in the most superior part of the upper extremity are the scapula (shoulder blade) and the clavicle (collar bone). These bones form the shoulder girdle and are the supporting foundation for the entire arm. The scapula is generally flat and triangular in shape and located at the posterior aspect of the shoulder. The scapula posseses a ridge into which is attached the trapezius muscle. The clavicle is located at the anterior aspect of the shoulder. The clavicle inserts into the sternum (breastbone) at one end and into the acromial process of the scapula at the other.

5.2.4.2 The Arm

The upper arm is composed of a single stout bone, the humerus. The proximal end of the humerus has a prominent semi-globular end which coincides with the glenoid cavity of the scapula. The distal end possesses two prominent protuberances which are the medial and lateral epicondyles. The forearm is composed of two bones: the radius and ulna, each smaller in size than the humerus. The ulna is the longer of the two bones. The proximal end of the ulna is the olecranon (elbow), which articulates closely with a depressed area of the humerus referred to as the trochlea. The major function of the ulna is to provide support to flexion and extension of the forearm in relationship to the upper arm. The other bone in the forearm is the radius. The radius articulates proximally to the semi-globular capitulum located at the distal end of the humerus. The major function of the radius is to support pronation-supination (rotation) of the forearm.

5.2.4.3 Wrist and Hand (Figure 5.6)

The wrist is composed of eight small carpal bones, generally triangular or rectangular in shape.

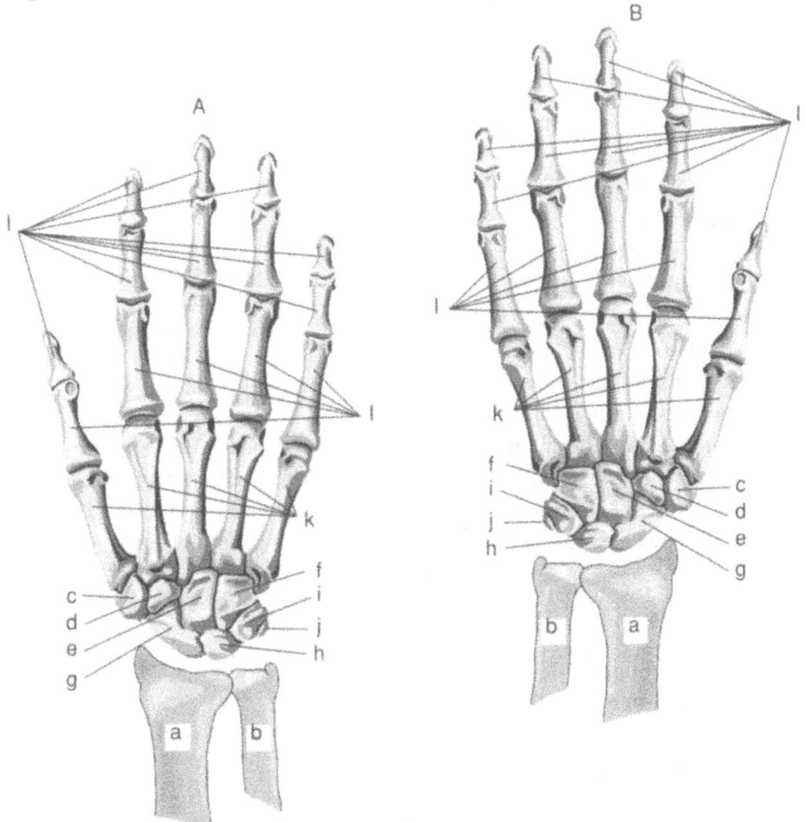

FIGURE 5.6 Bones of Wrist, Hand, Fingers

A. Posterior (dorsal) View
B. Anterior (palmar) View

a. Radius
b. Ulna
c. Trapezium
d. Trapezoid
e. Capitate
f. Hamate
g. Scaphoid
h. Lunate
i. Triquetrum
j. Pisiform
k. Metacarpals
l. Phalanges

The carpals are roughly arranged into two rows of four carpals each. The proximal row comprises the scaphoid, lunate, triquetrum, and pisiform bones. The scaphoid and lunate articulate with the distal end (styloid process) of the radius. The distal carpal row is composed of the capitate, hamate, trapezium, and trapezoid. These bones are intimately located next to the metacarpals and thumb. In general, there is some limited movement between the carpal bones. The carpal bones form a semi-circular ring (carpal tunnel) through which passes the median nerve and tendons for the fingers and thumb (Figure 5.7).

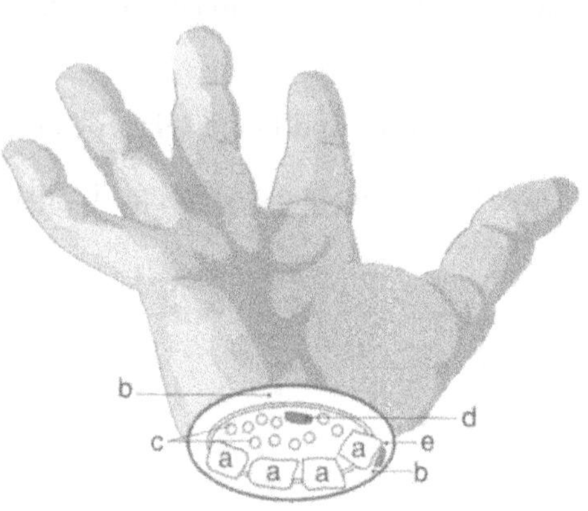

FIGURE 5.7 Cross Section of Carpal Tunnel

a. carpal bones
b. ligaments
c. tendons
d. Median Nerve
e. Radial Nerve

There are four metacarpals which form the palm of the hand and one metacarpal forming the base of the thumb. The distal ends of the metacarpals articulate with thumb and finger and form the knuckle joints. The four fingers are each composed of proximal, medial, and distal phalanges. There is articulation between the phalanges. Metacarpals and phalanges are numbered one to five beginning to the thumb.

5.3 MUSCLES

Muscles provide contractile forces to bones which move the bones and provide locomotion and motion for the organism.

5.3.1 Characteristics

There are three categories of muscles: 1) smooth, 2) cardiac, and 3) skeletal. Smooth muscles are involved with various visceral functions such as digestion. Smooth muscles, so named because of their appearance, are involuntary, in that the organism has no willful control over their implementation. Smooth muscles may be discovered in such areas as intestine and bladder. Cardiac muscle, found in the heart, is also involuntary and has the striped (striated) appearance of skeletal muscle. Skeletal muscles are called striated muscles because of their general appearance. They are the voluntary muscles which comprise the majority of the muscular system and provide the facilities for human movement. The cells of striated skeletal muscles are relatively elongated as compared to cells of other tissues and are frequently referred to as muscle fibers. A typical muscle has a tendon at either end. Tendons may be long or short in length and are frequently covered by a protective lubricating tendon sheath. The main body of the muscle, particularly at its widest cross-section, is called the belly.

The muscle tendon distal to the point of force is named the origin and acts as an anchorage between the muscle and the organism. The tendon proximal to the point of force is the insertion and transmits the muscle force to the moving bone. Muscles are generally covered in their entirety by a protective fascia. Beneath the fascia, the muscles are subdivided into muscle bundles which are further subdivided into fibers. Muscles are capable of performing their functions by contraction only. When a muscle is contracting and causing movement of a bone, it is referred to as a protagonist or prime mover. A muscle is generally paired off with another muscle which will expand and stretch the protagonist muscle after it has contracted. The other muscle is known as an antagonist. An example of the protagonist-antagonist relationship is the biceps muscles vs. the triceps muscle.

5.3.2 Major Muscles

There are a large number of muscles in the upper appendage, particularly in the forearm and hand. Some of the major muscles are herein listed. Their functions will be explained in the chapter on Kinesiology (Figures 5.8 to 5.14).

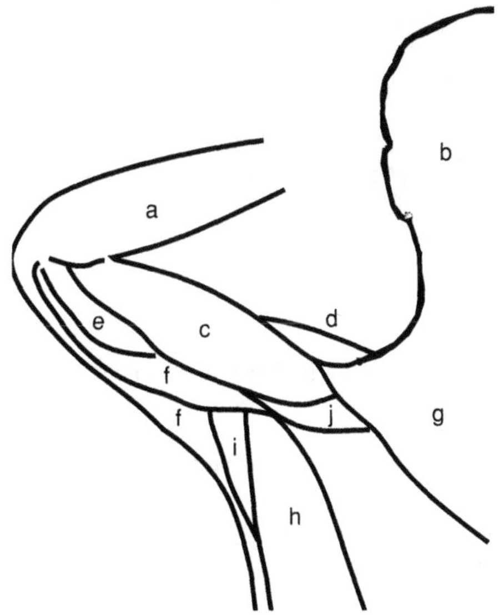

FIGURE 5.8 Anterior View, Shoulder and Upper Arm, Muscles

a. forearm	f. triceps
b. face	g. pectoralis
c. biceps	h. latissimus dorsi
d. deltoid	i. terres major
e. brachialis	j. coracobrachialis

5.3.2.1 Muscles Connecting Shoulder to Torso

These muscles support and hang the upper appendage from the axial skeleton.

Posterior View: Trapezius, rhomboideus major, rhomboideus minor, levator scapulae.
Anterior View: Pectoralis major, pectoralis minor, serratus anterior.

5.3.2.2 Muscles Connecting Arm to Torso

These muscles provide movement of the upper arm.

Posterior View: Latissimus dorsi.
Anterior View: Pectoralis major.

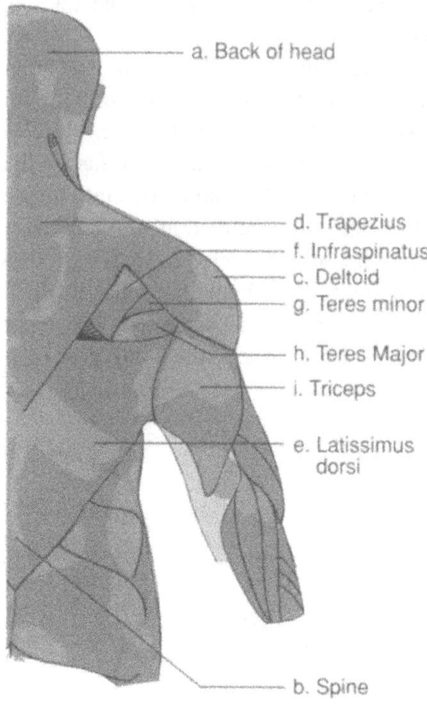

FIGURE 5.9 Posterior View of Shoulder, Muscles

a. back of head
b. spine
c. deltoid
d. trapezius
e. latissimus dorsi
f. infraspinatus
g. teres major
h. teres minor
i. triceps

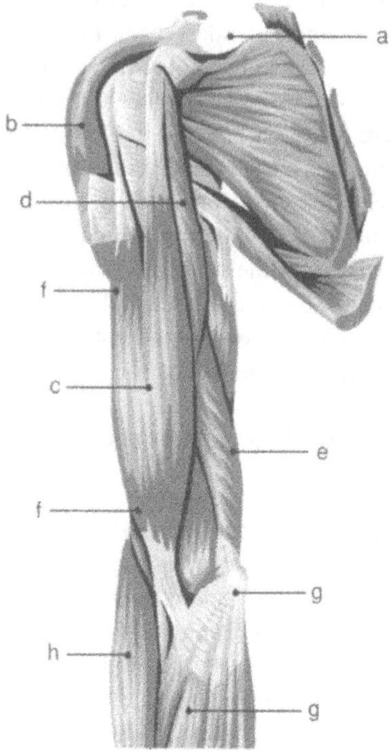

FIGURE 5.10 Anterior View, Shoulder and Upper Arm, Muscles

a. clavicle
b. deltoid
c. biceps
d. coracobrachialis
e. triceps
f. brachialis
g. ponator teres
h. brachioradialis

5.3.2.3 Muscles Connecting the Shoulder to Arm

These muscles provide movement of the upper arm.

Posterior View: Supraspinatus, infraspinatus
Anterior View: Deltoid, teres major, subscapularis.

5.3.2.4 Muscles of the Arm

These muscles provide movement of the forearm.

Posterior View: Triceps brachii.
Anterior View: Biceps brachii, brachialis.

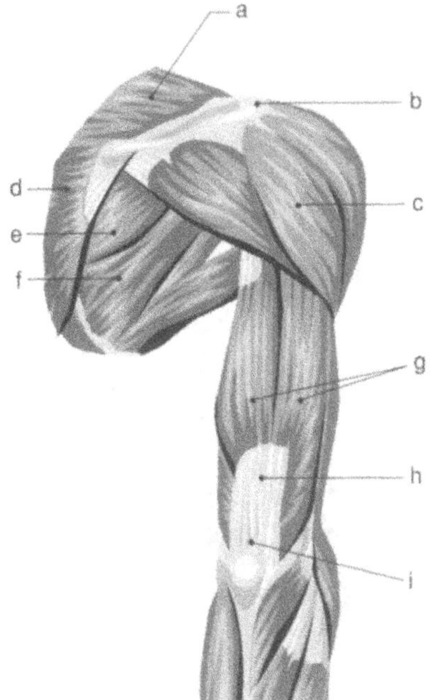

FIGURE 5.11 Posterior View, Shoulder and Upper Arm, Muscles

a. scapular ridge
b. acromion
c. deltoid
d. infraspinatus
e. teres minor
f. teres major
g. triceps
h. aponeurosis of triceps
i. olecranon

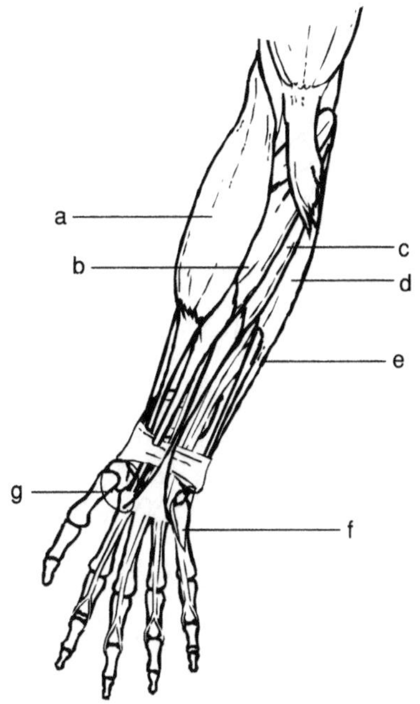

FIGURE 5.12 Anterior (Palmar) View Forearm, Muscles

a. brachioradialis
b. flexor carpi radialis
c. palmaris longus
d. flexor carpi ulnaris
e. tendons
f. hypothenar
g. thenar

5.3.2.5 Muscles of the Forearm

These muscles provide movement of forearm, hand, and finger. The large number of these muscles is best broken down into two major categories.

Posterior View: Extensor-supinator group.
Anterior View: Palmar flexor-pronator group.

5.3.2.6 Muscles within the Hand

These muscles provide movement of the fingers and thumb. Their large number is best broken down into three categories.

Thenar muscles which move the thumb.
Hypothenar muscles which move the smallest finger (fifth).
Intermediate muscles which move the last four fingers.

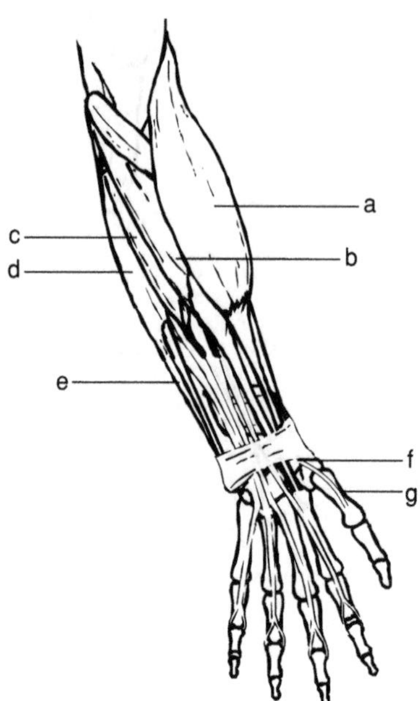

FIGURE 5.13 Posterior (Dorsal) View Forearm, Muscles

a. flexor carpi ulnaris
b. extensor carpi ulnaris
c. extensor digiti minimi
d. extensor digitorum communus
e. extensor carpi radialis brevis
f. extensor pollicis longus
g. tendons

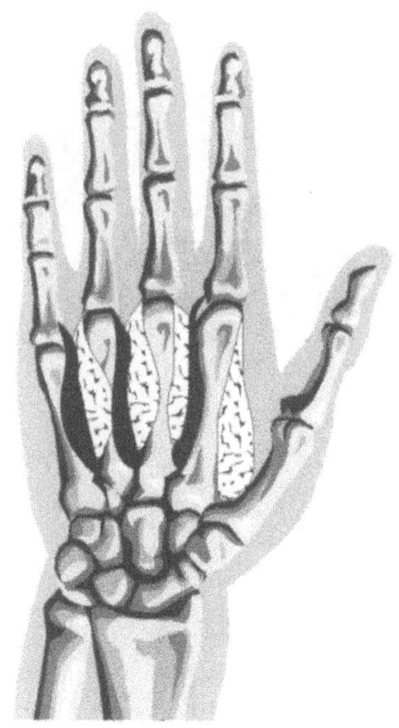

FIGURE 5.14 Palmar View, Muscles

The three palmar interossei muscles (black) adduct the fingers. The four dorsal interossei muscles (hatched) abduct the fingers.

5.4 CIRCULATORY SYSTEM

The circulatory system is predominantly composed of arteries and veins transporting blood to and away from tissue. Lymphatic vessels connect the arteries and the veins. Blood carried by the arteries contains oxygen, water, and nutritive materials needed for tissue function. Blood moved by the veins carries away carbon dioxide and other waste products from the tissues.

5.4.1 Major Arteries (Figures 5.15, 5.16)

The artery which descends from the torso is the axillary artery.

This artery becomes the brachial artery in the upper arm and then branches, in the forearm, into the radial artery in the vicinity of the radius bone and the ulnar artery in the vicinity of the ulna bone. The ulnar and radial arteries join together to form a palmar arch. Palmar digital arteries descend from the palmar arch into the palm and the four fingers.

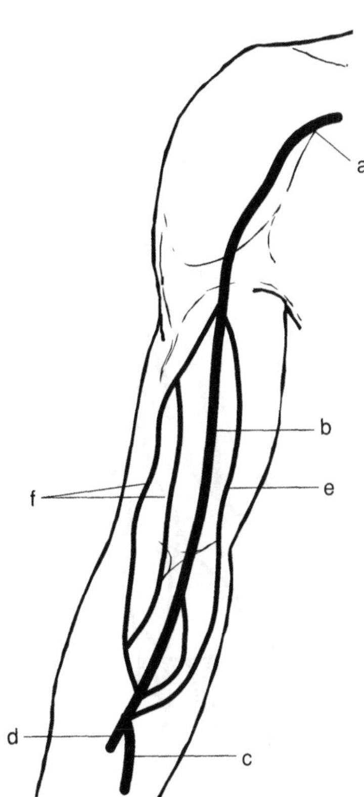

FIGURE 5.15 Upper Arm, Anterior View, Arteries

a. axillary
b. brachial
c. ulnar
d. radial
e. ulnar collateral
f. profunda brachii

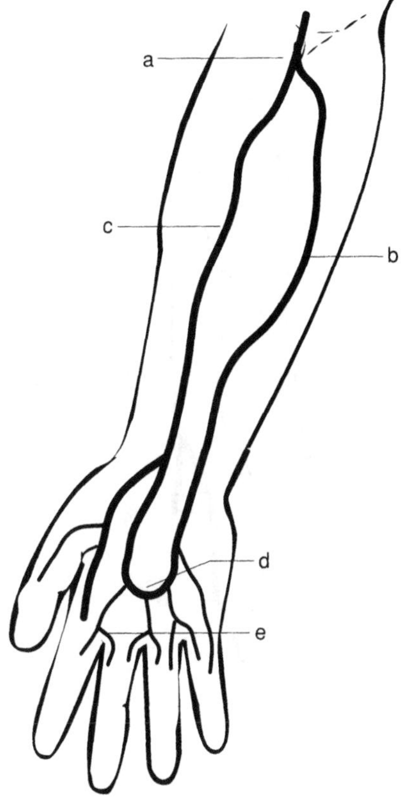

FIGURE 5.16 Lower Arm, Anterior View, Arteries

a. brachial
b. ulnar
c. radial
d. palmar arch
e. digital

5.4.2 Major Veins (Figures 5.17, 5.18)

The major veins of the arm follow and are analogous to the major arteries. These veins are the axillary vein, brachial vein, radial vein, and ulnar vein. In addition, there are two long continuous veins extending from hand to shoulder. These veins are the cephalic vein located at the lateral aspect of the arm and the basilic vein located at the medial aspect of the arm.

5.5 NERVOUS SYSTEM (FIGURE 5.19)

The most important basic functions of nerves are reception and transmission. Nerves, which transmit sensory messages to spinal cord and brain, have an afferent function. Nerves, which transmit locomotive messages to the muscles, have efferent functions.

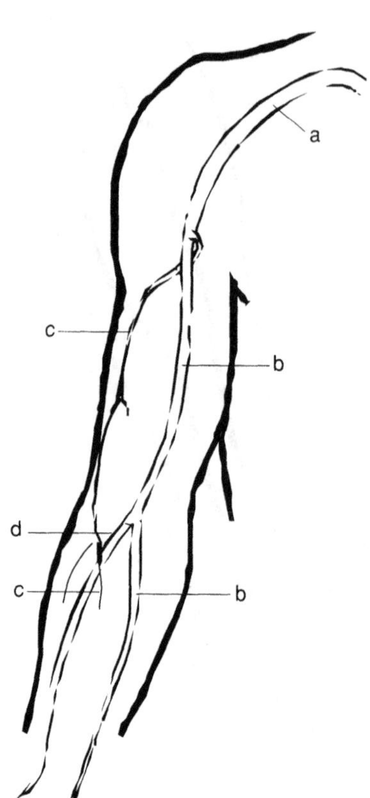

FIGURE 5.17 Upper Arm, Anterior View, Veins

a. axillary
b. basilic
c. cephalic
d. deep median

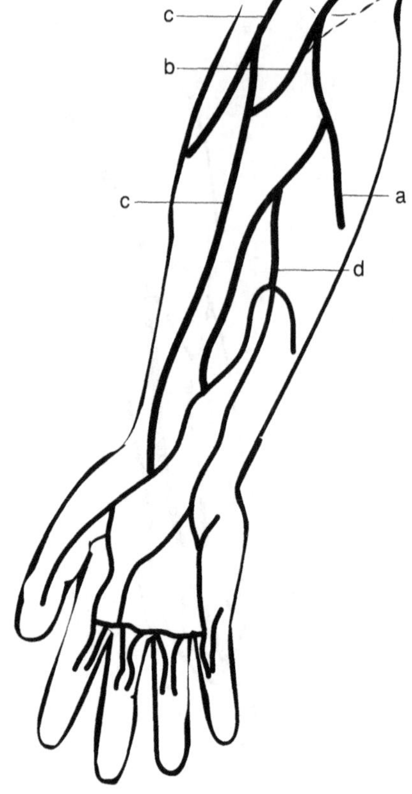

FIGURE 5.18 Lower Arm, Anterior View, Veins

a. basilic
b. median cubital
c. cephalic
d. median

5.5.1 Major Nerves

The major nerves of arm and hand are considered peripheral nerves. They emanate from the fifth, sixth and seventh cervical and first thoracic vertebrae of the spine. There are three major nerves supplying efferent and afferent functions in arm and hand: 1) the radial nerve extends the entire length of the arm and hand. The radial nerve begins beneath the humerus (anterior view), then moves downward above the radius and then ends below the radius. 2) the ulnar nerve moves downward medially to the humerus (anterior view), passes beneath the medial epicondyle, continues medially to the ulna and ends (innervates) at the fourth and fifth digits. 3) the medial nerve moves downward generally in between the ulnar and radial nerves, passes through the carpal tunnel and ends (innervates) at the first four digits.

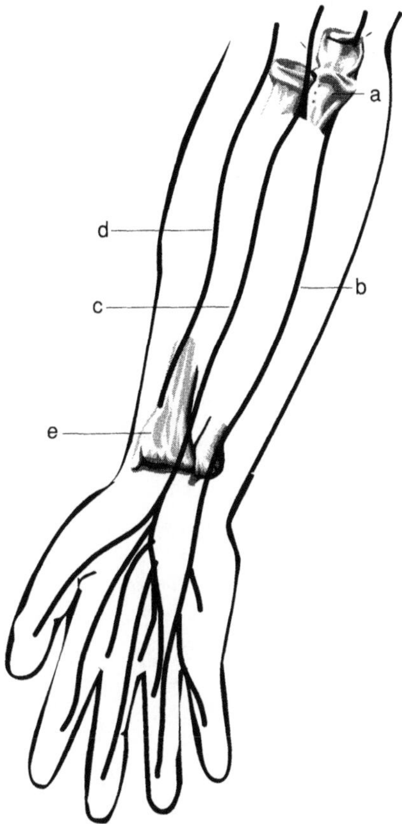

FIGURE 5.19 Lower Arm, Anterior View, Nerves

a. end of humerus
b. ulnar nerve
c. median nerve
d. radial nerve
e. radius

6 Anthropometry of the Hand

6.1 DISCUSSION

In essence, anthropometry is the discipline which executes, records, and analyzes various measurements related to the external aspects of the human anatomy. In addition to anatomical measurements, anthropometry also measures the mass (weight) of human beings, as well as human capabilities at exerting force (strength). Pheasant (1986) indicates that anthropometry had its formalized beginning with the painter Albrecht Durer (1471-1528). Since that time, anthropometry has developed into a discipline which has created a large bank of information about the topographies and typologies of people.

6.2 GOAL

The ultimate utilitarian goal of anthropometry is to supply human measurements which will optimize the physical interface of the human being, and any artifact which the human is utilizing. It is generally an accepted fact that an optimal "fit" of human and vehicle, furniture, edifice, machine control, or implement will: 1) minimize errors in use, and 2) minimize biomechanical stresses. These minimizations, in turn, will lead to 1) greater productivity, 2) greater efficiency, 3) fewer mishaps leading to injury, and property damage, and 4) fewer musculoskeletal diseases and other diseases as well.

6.3 IMPORTANT CHARACTERISTICS

Anthropometrists work under a number of common protocols which are based upon logical scientific research procedures.

6.3.1 Large Numbers

Measurements are executed upon large numbers of humans. Measurements of only a few individuals would be considered unsatisfactory. Mass measurements, which may include as many as 1,000 individuals, are intended to describe a "typicality" of humanity in general or of large segments and subsegments of humanity.

6.3.2 Procedures in Measurement

To prevent "apples vs. oranges" comparisons, all members of the group being measured are measured in the same manner. Kroemer, Kroemer, and Kroemer-Elbert

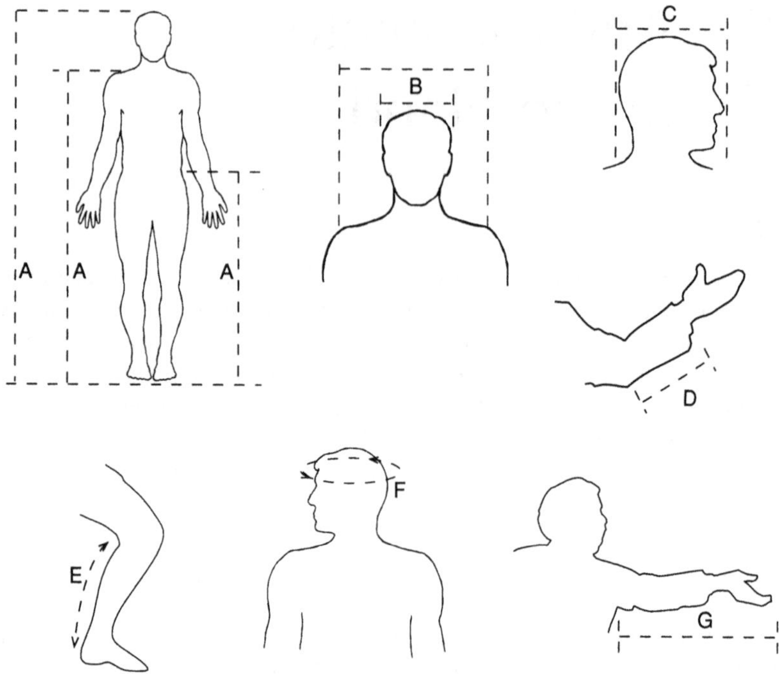

FIGURE 6.1 Measurements.

A. Height
B. Breadth
C. Depth
D. Length
E. Curvature
F. Circumference
G. Reach

(1990) describe some of the most standardized measurement techniques. Measurements which are usually executed are (Figure 6.1):

Height. A straight line point-to-point vertical measurement.
Breadth. A straight line point-to-point horizontal measurement.
Depth. A straight line point-to-point forward and back measurement.
Distance. A straight line point-to-point measurement between anatomical landmarks.
Curvature. A point-to-point measurement along an open anatomical curve (convex or concave).
Circumference. A point-to-point measurement around an anatomical contour, usually not circular.
Reach. A point measurement following the long axis of arm and leg.

Measurement devices which are commonly used:

Measuring Tape. A flexible rule on which is imprinted gradation in measurements.

Calipers and Anthropometer. Establishes the span between two anatomical points. The span is then measured (Figure 6.2).

Other Methods. In addition to the traditional methods above, grids marked upon walls have been used as well as photographic techniques.

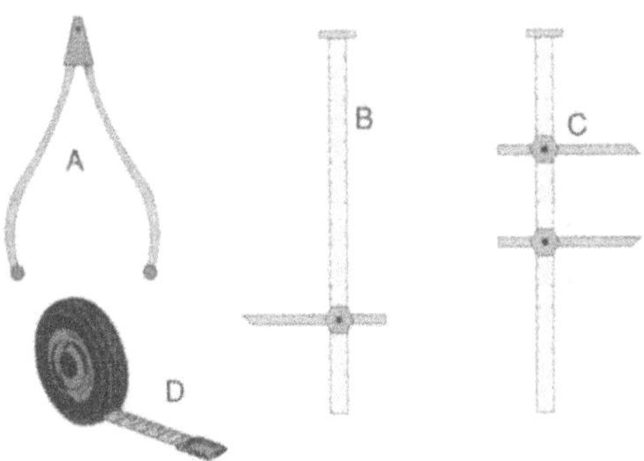

FIGURE 6.2 Measuring Instruments.

A. Curved calipers C. Straight calipers
B. Anthropometer D. Measuring tape

6.3.3 Reporting and Analyzing Measurements

The various measurements based upon large numbers of individuals are generally reported in tables. Measurements are reported in the english system or the metric system or both. The various measurements are normally distributed, thus, parametric statistics are used to analyze the data. A table minimally reports, for specific measurements, percentiles along the curve of normal distribution. A percentile is a ranking number which indicates the percentage of individuals which manifests a particular dimension or less. If a table indicates that the 95th percentile of a group is 183 centimeters in stature (height), the inference is that 95% of those measured are at least 183 centimeters tall and 5% are over 183 centimeters tall. The percentiles most frequently reported are the 5th, 50th, and 95th. Other percentiles may also be reported as well. The mean of the measurement in question, because of normality in distribution, is very similar to the fiftieth percentile. The standard deviation for a specific measurement may also be provided so that any percentile along the curve can be calculated.

6.3.4 Subgroups

Although anthropometry may be an attempt at reporting dimensions about people in general, subgroups of humanity may be reported because of specific needs or requirements. Thus, separate tables or sections of tables will give separate dimensions of males and females. There may also be separate reporting of racial or ethnic groups or professional/occupational groups.

6.4 APPLYING THE INFORMATION

Designers of artifacts who use anthropomorphic tables will make use of the tables in accordance with unique design requirements. Pheasant (1986) and Sanders and McCormick (1993) describe various design decisions made under various design circumstances.

6.4.1 Clearance

Sometimes individuals must pass through or fit into apertures. In these cases, a designer will specify an artifact's dimensions so that the aperture accommodates the size of the largest individuals, knowing that this accommodation will also accommodate the size of the smallest individuals. The classic example is specifying a doorway height of 7 feet. This height will accommodate an exceptionally tall person but will also accommodate an exceptionally short person. Usually, dimensions using this principle will be based upon the dimensions of a 95th percentile male.

6.4.2 Reach

Sometimes individuals must extend parts of themselves outward from their torso in order to acquire objects, place objects, or perform manipulations. In these cases, a designer will specify an artifact's dimensions so that the artifact accommodates the reach of the smallest individuals, knowing that this accommodation will also accommodate the largest individuals. The classic example is specifying a short distance between a chair and a control panel. This short reach will serve the small person and the large person as well. Usually dimensions using this principle will be based upon the dimensions of fifth percentile females.

6.4.3 Adjustability

Some artifacts may be so designed that their critical dimensions can be varied by adjustment. This principle, which is frequently applied to furniture, allows the small user or large user to make dimensional adjustments to suit his/her own needs related to fit. An example of this principle application is tables and chairs of adjustable height. Usually the range of adjustability will vary from the dimensions of a 5th percentile female to a 95th percentile male. This adjustability range will serve approximately 95% of all who might use the table and chair.

6.4.4 Average Dimension

In some instances, because of functional, procedural, or economic reasons, the principle of adjustability cannot be applied. In these cases the "average human" dimension (50th percentile of males and females combined) is utilized. An example of this principle is the height of a checkout counter in a supermarket.

6.4.5 Adaptability

The aforementioned principles will apply to the largest majority of individuals. In cases of extremities in size (Lilliputian or Brobdingnagian) such individuals: 1) are excluded from the use of the artifact, or 2) are provided with custom-made artifacts suited to their dimensions, or 3) use the standard artifact provided but with some inconvenience. Examples: the giant will stoop when walking through a doorway, a little person will place a telephone book on the seat of his/her chair. In the case of the average-dimensioned artifact, such as the supermarket checkout counter, the inconvenience also affects those who are not extreme in size, but the inconvenience may not be very great.

6.5 CORRELATION AND ADDITIVITY

Anthropometric tables may also provide correlation coefficients which indicate whether there is a strong or weak relationship between the dimensions of various parts of the anatomy. As would be expected, there is a high positive correlation between stature and acromion (shoulder) height. Kroemer, et al., indicate that some guarded predictions can be made if only one dimension is known, provided that the correlation is at least +.7. Thus, knowing shoulder height and knowing head length, a designer can reasonably add the two dimensions to determine stature provided there is at least a +.7 correlation between the two segments.

6.6 TABLES

The appendix at the end of this chapter contains an extensive assortment of measurements of the hand and arm (Table 6.1, Table 6.2, Figures 6.3A, 6.3B, 6.3C, 6.3D, 6.3E, 6.3F, 6.3G, 6.3H, 6.3I, 6.3J). The measurements are extracted from a U.S. Army report. Measurements supplied are at the 5th, 50th, and 95th percentiles plus means and standard deviations.

TABLE 6.1
Hand and Arm Dimensions — Male

	Percentile						Standard Deviation		Mean	
	5th		50th		95th					
Dimension #	cm	in.	cm	in.	cm	in.	cm	in.	cm	in.
1	6.20	2.44	6.97	2.74	7.75	3.05	.48	.19	6.97	2.74
2	8.82	3.47	10.03	3.95	11.24	4.43	.74	.29	10.03	3.95
3	12.42	4.89	13.75	5.41	15.28	6.02	.87	.34	13.79	5.43
4	2.19	.86	2.39	.94	2.64	1.04	.13	.05	2.40	.94
5	6.79	2.67	7.17	2.82	7.75	3.05	.29	.12	7.23	2.85
6	11.20	4.41	12.31	4.85	13.57	5.34	.72	.28	12.34	4.86
7	7.15	2.81	8.18	3.22	9.45	3.72	.71	.28	8.23	3.24
8	1.62	.64	2.11	.83	2.62	1.03	.30	.12	2.11	.83
9	3.03	1.19	3.44	1.35	3.90	1.53	.26	.10	3.45	1.36
10	6.74	2.65	7.52	2.96	8.35	3.29	.49	.19	7.53	2.96
11	16.51	6.50	17.98	7.08	19.60	7.72	.95	.37	18.00	7.09
12	16.98	6.68	18.48	7.28	20.17	7.94	.99	.39	18.52	7.29
13	2.04	.80	2.29	.90	2.58	1.02	.16	.06	2.30	.90
14	6.56	2.58	6.83	2.69	7.15	2.82	.18	.07	6.84	2.69
15	1.77	.70	2.00	.79	2.28	.90	.15	.06	2.01	.79
16	5.48	2.16	5.72	2.25	6.03	2.37	.16	.06	5.74	2.26
17	9.75	3.84	10.81	4.25	12.01	4.73	.69	.27	10.83	4.26
18	6.96	2.74	7.66	3.02	8.50	3.34	.48	.19	7.68	3.02
19	2.45	.96	2.84	1.12	3.22	1.27	.23	.09	2.84	1.12
20	1.87	.74	2.25	.88	2.68	1.06	.24	.10	2.26	.89
21	5.18	2.04	6.05	2.38	7.08	2.79	.59	.23	6.08	2.39
22	7.52	2.96	8.37	3.29	9.29	3.66	.54	.21	8.38	3.30
23	17.85	7.03	19.36	7.62	21.17	8.33	1.03	.40	19.41	7.64
24	17.87	7.03	19.40	7.64	21.20	8.35	1.03	.41	19.45	7.66
25	1.99	.79	2.24	.88	2.52	.99	.16	.06	2.25	.88
26	6.64	2.61	6.95	2.74	7.31	2.88	.20	.08	6.96	2.74
27	1.74	.69	1.97	.78	2.23	.88	.14	.06	1.98	.78
28	5.54	2.18	5.77	2.27	6.05	2.38	.16	.06	5.78	2.28
29	9.88	3.89	10.97	4.32	12.16	4.79	.70	.28	10.99	4.32
30	7.49	2.95	8.45	3.33	9.49	3.74	.60	.24	8.46	3.33
31	2.45	.96	2.84	1.12	3.22	1.27	.23	.09	2.84	1.12
32	2.19	.86	2.62	1.03	3.13	1.23	.28	.11	2.64	1.04
33	4.71	1.85	5.46	2.15	6.33	2.49	.50	.20	5.48	2.16
34	7.09	2.79	7.90	3.11	8.80	3.46	.52	.21	7.92	3.12
35	16.51	6.50	17.95	7.07	19.78	7.79	1.01	.40	18.02	7.09
36	16.98	6.68	18.43	7.26	20.27	7.98	1.02	.40	18.50	7.28
37	1.91	.75	2.13	.84	2.41	.95	.15	.06	2.14	.84
38	6.19	2.44	6.47	2.55	6.83	2.69	.19	.07	6.49	2.55
39	1.62	.64	1.84	.72	2.10	.83	.14	.06	1.85	.73

TABLE 6.1 (continued)
Hand and Arm Dimensions — Male

	Percentile						Standard Deviation		Mean	
	5th		50th		95th					
Dimension #	cm	in.	cm	in.	cm	in.	cm	in.	cm	in.
40	5.15	2.03	5.37	2.11	5.62	2.21	.13	.05	5.38	2.12
41	9.67	3.81	10.67	4.20	11.77	4.63	.65	.25	10.69	4.21
42	6.92	2.72	7.79	3.07	8.79	3.46	.56	.22	7.81	3.08
43	2.56	1.01	2.97	1.17	3.35	1.32	.24	.10	2.96	1.17
44	2.01	.79	2.42	.95	2.88	1.13	.26	.10	2.43	.96
45	4.66	1.84	5.28	2.08	5.96	2.35	.40	.16	5.29	2.08
46	5.66	2.23	6.47	2.55	7.28	2.87	.49	.19	6.47	2.55
47	13.07	5.15	14.50	5.71	16.14	6.36	.94	.37	14.54	5.72
48	14.47	5.70	15.94	6.27	17.69	6.97	.98	.38	15.99	6.30
49	1.71	.67	1.91	.75	2.14	.84	.13	.05	1.92	.75
50	5.50	2.16	5.78	2.28	6.11	2.41	.18	.07	5.78	2.28
51	1.53	.60	1.73	.68	1.96	.77	.13	.05	1.74	.68
52	4.68	1.84	4.91	1.93	5.18	2.04	.16	.06	4.92	1.94
53	7.65	3.01	8.58	3.38	9.60	3.78	.59	.23	8.60	3.39
54	6.45	2.54	7.37	2.90	8.41	3.31	.60	.24	7.39	2.91
55	2.35	.93	2.72	1.07	3.12	1.23	.23	.09	2.73	1.07
56	1.37	.54	1.75	.69	2.12	.83	.22	.09	1.75	.69
57	3.59	1.41	4.14	1.63	4.75	1.87	.36	.14	4.15	1.63
58	17.85	7.03	19.36	7.62	21.17	8.33	1.03	.40	19.41	7.64
59	17.85	7.03	19.35	7.62	21.09	8.30	.99	.39	19.41	7.64
60	19.83	7.81	21.36	8.41	23.05	9.07	.98	.39	21.39	8.42
61	10.13	3.99	11.02	4.34	12.08	4.75	.60	.23	11.05	4.35
62	8.61	3.39	9.52	3.75	10.50	4.13	.58	.23	9.53	3.75
63	8.36	3.29	9.02	3.55	9.76	3.84	.42	.17	9.04	3.56
64	5.87	2.31	6.56	2.58	7.35	2.89	.45	.18	6.58	2.59
65	16.17	6.36	17.40	6.85	18.84	7.42	.82	.32	17.43	6.86
66	6.25	2.46	6.95	2.73	7.83	3.08	.48	.19	6.98	2.75
67	16.66	6.56	18.07	7.12	19.65	7.74	.91	.36	18.10	7.13
68	11.36	4.47	12.43	4.89	13.61	5.36	.68	.27	12.45	4.90
69	6.16	2.43	6.89	2.71	7.72	3.04	.47	.19	6.91	2.72
70	10.11	3.98	11.02	4.34	12.07	4.75	.61	.24	11.04	4.35
71	10.01	3.94	10.96	4.32	12.09	4.76	.64	.25	10.99	4.33
72	8.71	3.43	9.62	3.79	10.73	4.22	.62	.24	9.66	3.80
73	44.82	17.65	48.24	18.99	52.46	20.65	2.33	.92	48.40	19.06
74	26.65	10.49	28.92	11.38	31.64	12.46	1.53	.60	29.00	11.42
75	33.34	13.13	35.83	14.11	39.09	15.39	1.77	.70	35.98	14.16

Extracted from a U.S. Army report.

TABLE 6.2
Hand and Arm Dimensions — Female

	Percentile						Standard Deviation		Mean	
	5th		50th		95th					
Dimension #	cm	in.	cm	in.	cm	in.	cm	in.	cm	in.
1	5.58	2.20	6.33	2.49	7.17	2.82	.48	.19	6.35	2.50
2	8.07	3.18	9.25	3.64	10.47	4.12	.73	.29	9.26	3.65
3	11.19	4.41	12.53	4.93	14.08	5.54	.87	.34	12.57	4.95
4	1.86	.73	2.06	.81	2.29	.90	.13	.05	2.06	.81
5	5.93	2.33	6.31	2.49	6.75	2.66	.25	.10	6.30	2.48
6	9.94	3.91	11.04	4.34	12.22	4.81	.69	.27	11.05	4.35
7	6.50	2.56	7.52	2.96	8.81	3.47	.70	.28	7.57	2.98
8	1.47	.58	1.92	.75	2.41	.95	.29	.11	1.92	.76
9	2.69	1.06	3.07	1.21	3.49	1.37	.24	.10	3.08	1.21
10	6.19	2.44	6.95	2.74	7.73	3.04	.46	.18	6.96	2.74
11	15.06	5.93	16.49	6.49	18.04	7.10	.90	.36	16.51	6.50
12	15.50	6.10	16.95	6.67	18.59	7.32	.93	.37	16.99	6.69
13	1.78	.70	1.98	.78	2.20	.87	.13	.05	1.99	.78
14	5.78	2.27	6.11	2.40	6.49	2.56	.20	.08	6.12	2.41
15	1.54	.61	1.72	.68	1.94	.77	.12	.05	1.73	.68
16	4.77	1.88	5.09	2.00	5.40	2.12	.19	.07	5.08	2.00
17	9.00	3.54	10.01	3.94	11.12	4.38	.64	.25	10.02	3.95
18	6.22	2.45	6.93	2.73	7.82	3.08	.49	.19	6.96	2.74
19	2.20	.87	2.55	1.00	2.90	1.74	.21	.08	2.55	1.00
20	1.77	.70	2.10	.83	2.50	.98	.22	.09	2.11	.83
21	4.87	1.92	2.60	2.20	6.62	2.61	.53	.21	5.65	2.22
22	6.91	2.72	7.71	3.03	8.59	3.38	.51	.20	7.72	3.04
23	16.26	6.40	17.75	6.99	19.48	7.67	.98	.39	17.79	7.00
24	16.31	6.42	17.79	7.01	19.53	7.69	.98	.39	17.84	7.02
25	1.73	.68	1.93	.76	2.15	.85	.13	.05	1.93	.76
26	5.83	2.29	6.13	2.41	6.47	2.55	.19	.07	6.13	2.41
27	1.52	.60	1.71	.67	1.91	.75	.11	.05	1.71	.67
28	4.82	1.90	5.10	2.01	5.38	2.12	.17	.07	5.09	2.00
29	9.02	3.55	10.01	3.94	11.11	4.37	.64	.25	10.03	3.95
30	6.96	2.74	7.77	3.06	8.78	3.46	.56	.22	7.81	3.07
31	2.20	.87	2.55	1.00	2.90	1.14	.21	.08	2.55	1.00
32	2.09	.82	2.49	.98	2.99	1.18	.27	.11	2.51	.99
33	4.29	1.69	4.95	1.95	5.72	2.25	.44	.17	4.97	1.96
34	6.42	2.53	7.21	2.84	8.06	3.17	.50	.20	7.22	2.84
35	14.97	5.89	16.42	6.47	18.09	7.12	.96	.38	16.46	6.48
36	15.38	6.06	16.84	6.63	18.59	7.32	.98	.38	16.89	6.65
37	1.64	.65	1.83	.72	2.06	.81	.12	.05	1.84	.72
38	5.43	2.14	5.72	2.25	6.08	2.39	.19	.08	5.74	2.26
39	1.39	.55	1.57	.62	1.78	.70	.11	.05	1.58	.62

TABLE 6.2 (continued)
Hand and Arm Dimensions — Female

	Percentile						Standard Deviation		Mean	
	5th		50th		95th					
Dimension #	cm	in.	cm	in.	cm	in.	cm	in.	cm	in.
40	4.41	1.74	4.67	1.84	4.95	1.95	.16	.06	4.68	1.84
41	8.80	3.46	9.71	3.82	10.74	4.23	.59	.23	9.73	3.83
42	6.34	2.50	7.13	2.81	8.10	3.19	.53	.21	7.16	2.82
43	2.25	.89	2.60	1.02	2.98	1.17	.22	.09	2.61	1.03
44	1.87	.74	2.26	.89	2.74	1.08	.26	.10	2.28	.90
45	4.31	1.70	4.83	1.90	5.43	2.14	.34	.13	4.84	1.91
46	5.08	2.00	5.83	2.30	6.58	2.59	.46	.18	5.83	2.30
47	11.79	4.64	13.19	5.19	14.68	5.78	.88	.35	13.21	5.20
48	13.09	5.15	14.51	5.71	16.19	6.38	.94	.37	14.55	5.73
49	1.46	.58	1.65	.65	1.85	.73	.11	.04	1.56	.65
50	4.78	1.88	5.05	1.99	5.36	2.11	.17	.07	5.06	1.99
51	1.29	.51	1.47	.58	1.66	.65	.11	.04	1.47	.58
52	4.01	1.58	4.24	1.67	4.51	1.78	.15	.06	4.25	1.67
53	6.89	2.71	7.75	3.05	8.65	3.41	.54	.21	7.76	3.05
54	5.86	2.31	6.78	2.67	7.79	3.07	.59	.23	6.79	2.67
55	2.03	.80	2.37	.93	2.71	1.07	.20	.08	2.37	.93
56	1.29	.51	1.62	.64	2.00	.79	.22	.09	1.63	.64
57	3.28	1.29	3.78	1.49	4.28	1.68	.30	.12	3.78	1.49
58	16.26	6.40	17.75	6.99	19.48	7.67	.98	.39	17.79	7.00
59	16.53	6.51	18.03	7.10	19.76	7.78	.98	.39	18.07	7.11
60	17.28	6.30	18.63	7.33	20.09	7.91	.86	.34	18.65	7.34
61	9.20	3.62	10.07	3.96	11.05	4.35	.57	.22	10.09	3.97
62	7.59	2.99	8.31	3.27	9.04	3.56	.44	.17	8.31	3.27
63	7.35	2.90	7.94	3.13	8.59	3.38	.38	.15	7.95	3.13
64	5.15	2.03	5.69	2.24	6.27	2.47	.34	.14	5.70	2.24
65	14.05	5.53	15.11	5.95	16.32	6.43	.69	.27	15.14	5.96
66	5.87	2.31	6.59	2.59	7.49	2.95	.49	.19	6.62	2.61
67	15.49	6.10	16.93	6.67	18.46	7.27	.90	.35	16.95	6.67
68	10.69	4.21	11.75	4.63	12.92	5.09	.68	.27	11.77	4.63
69	5.56	2.19	6.26	2.46	7.07	2.78	.46	.18	6.28	2.47
70	9.15	3.60	10.03	3.95	11.04	4.35	.57	.23	10.05	3.96
71	9.11	3.59	10.00	3.94	11.08	4.36	.60	.24	10.03	3.95
72	7.88	3.10	8.77	3.45	9.78	3.85	.58	.23	8.79	3.46
73	40.61	15.99	44.28	17.43	48.33	19.03	2.36	.93	44.35	17.46
74	23.78	9.36	26.26	10.34	28.86	11.36	1.56	.61	26.28	10.35
75	30.01	11.81	32.87	12.94	35.85	14.12	1.78	.70	32.90	12.95

Extracted from a U.S. Army report.

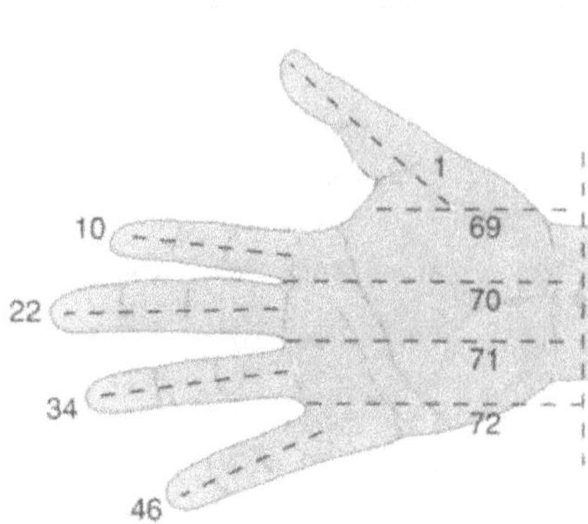

FIGURE 6.3A Right Palm. Key to dimensions.

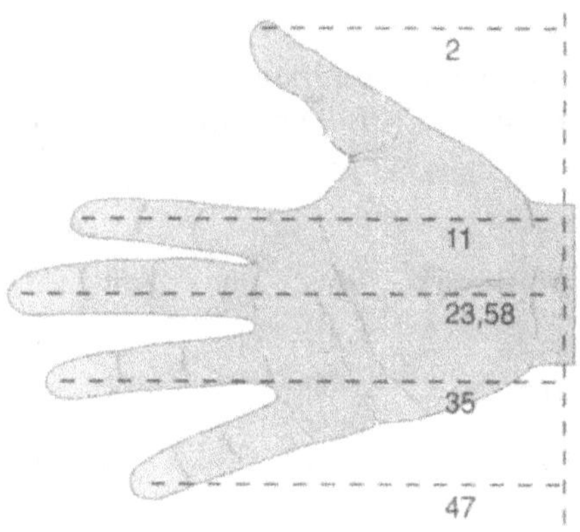

FIGURE 6.3B Right Palm. Key to dimensions.

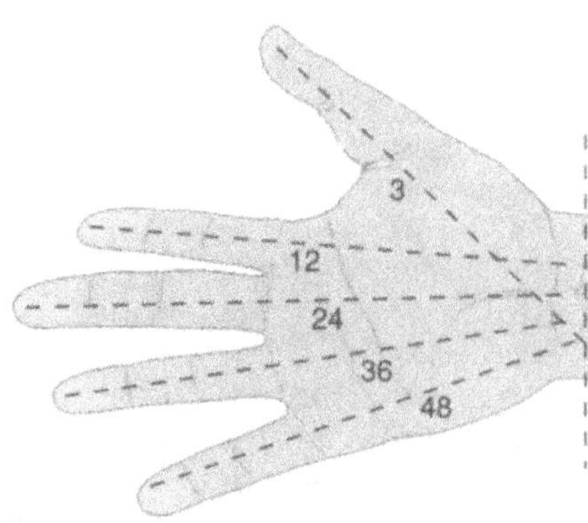

FIGURE 6.3C Right Palm. Key to dimensions.

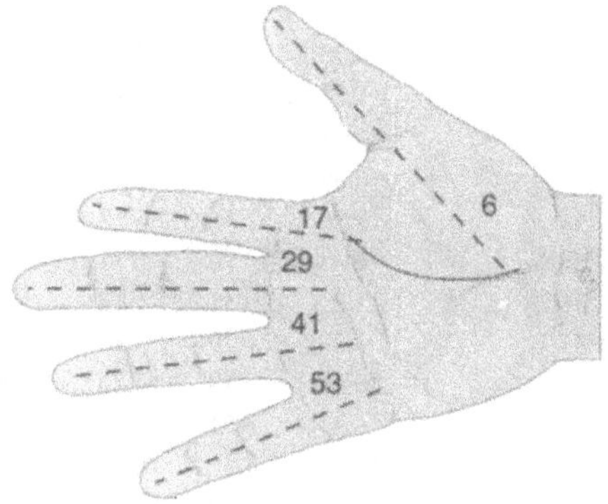

FIGURE 6.3D Right Palm. Key to dimensions.

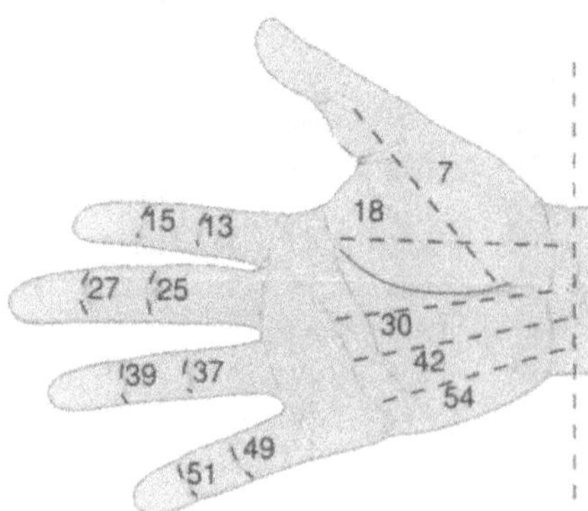

FIGURE 6.3E Right Palm. Key to dimensions.

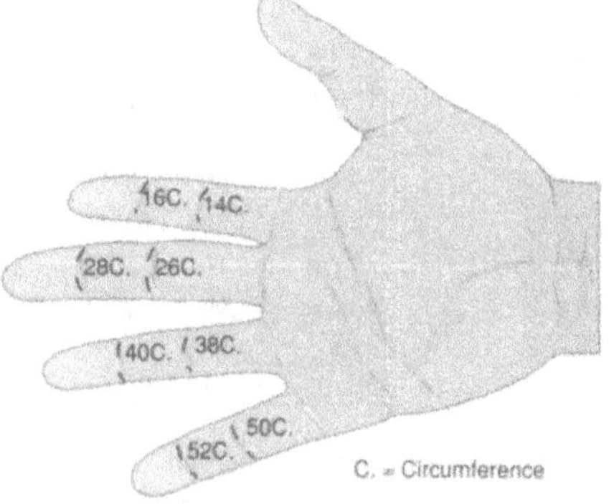

FIGURE 6.3F Right Palm. Key to dimensions.

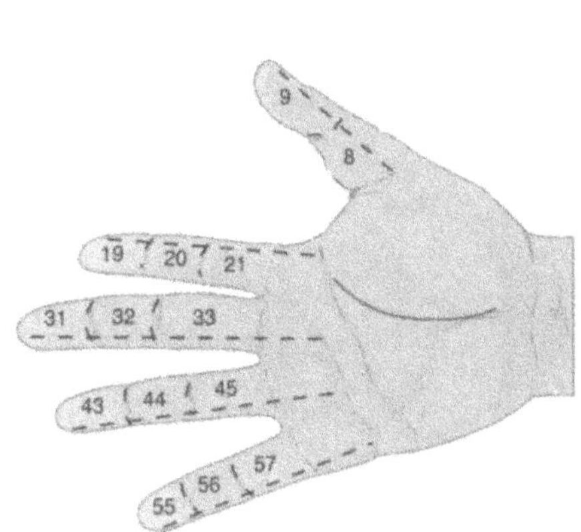

FIGURE 6.3G Right Palm. Key to dimensions.

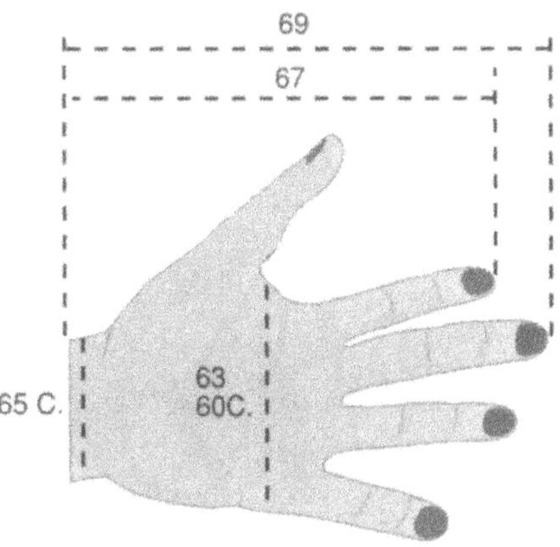

FIGURE 6.3H Back of Right Hand. Key to dimensions.

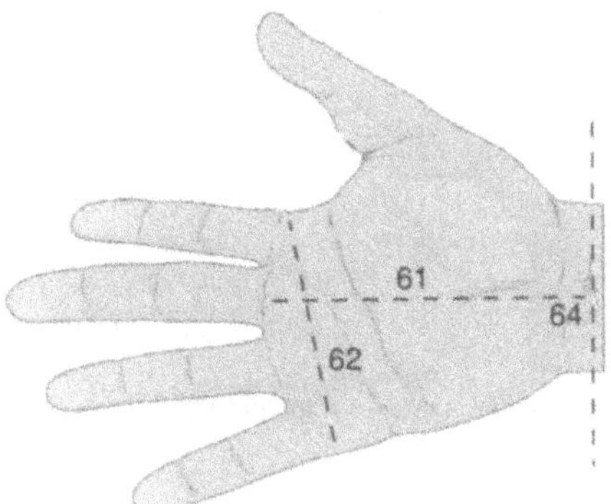

FIGURE 6.3I Right Palm. Key to dimensions.

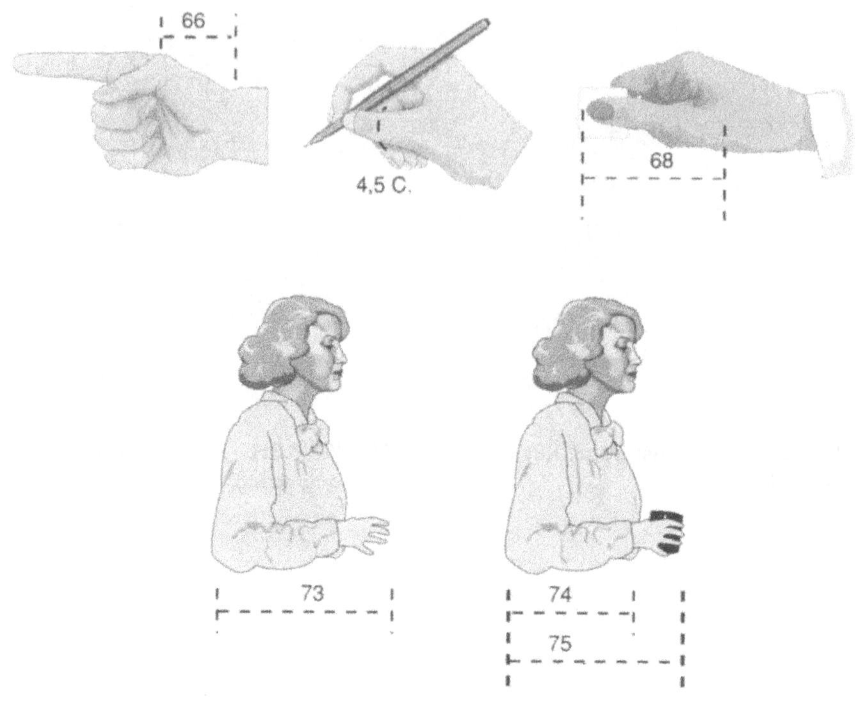

FIGURE 6.3J Various Aspects. Key to dimensions.

7 Kinesiology of Arm and Hand

7.1 INTRODUCTION

Chapter 5 (Anatomy) gives us information about the structural framework and the static aspects of hand and arm. This chapter, dealing with kinesiology, provides information about the dynamic use of the upper limb.

7.2 KINESIOLOGY

Kinesiology is the discipline which studies the motions of animals in their entirety or studies the motions of individual animal segments. This section will deal with: 1) movements within human joints, 2) basic human movements, 3) specific movements of hand and arm, and 4) manipulations of the hand.

7.2.1 Movements within Joints

Movements within joints are dependent upon the structure and configuration of the joint in question. As already indicated in the chapter on anatomy, diarthrodial joints comprise the majority of the joints involved in human motion. There are several categories of diarthrodial joints.

7.2.1.1 Hinge Joint (Figure 7.1)

A joint analogous to a door hinge. Only one direction of movement, such as flexion-extension, is possible. Because of unidirectionality, this joint is considered to have only 1 degree of freedom. An example of a hinge joint is the elbow.

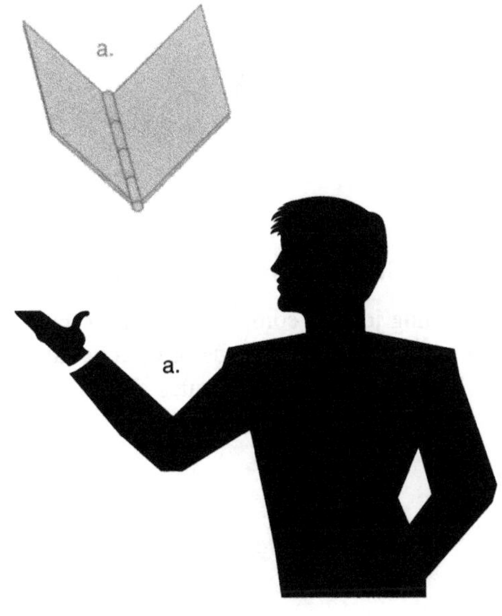

FIGURE 7.1 a. Hinge joint, elbow

7.2.1.2 Pivot Joint (Figure 7.2)

A joint which allows rotation in a single axis. This joint, because it has motion in one dimension, has 1 degree of freedom. An example of a pivot joint is the head rotating at the atlas joint at the top of the spine, nodding the "no" gesture.

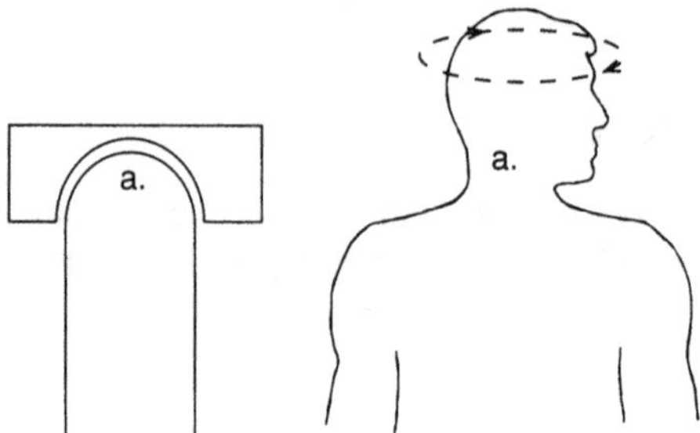

FIGURE 7.2 a. Pivot joint, base of skull

7.2.1.3 Ball-and-Socket Joint (Figure 7.3)

These joints display the greatest latitude of movement. They have 3 degrees of freedom and comprise a ball-shaped head inserted into a socket. The hip is an example of a ball-and-socket joint.

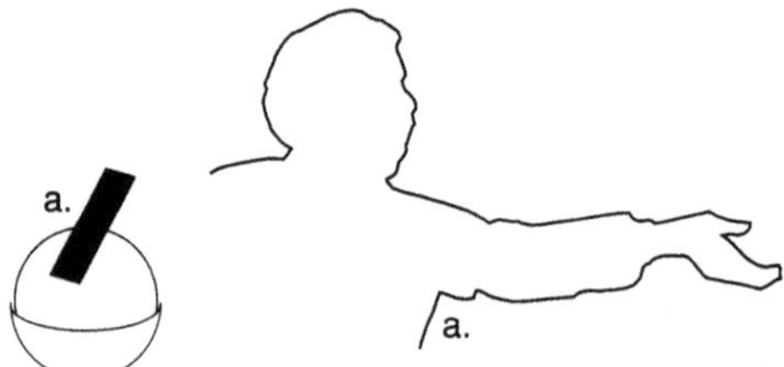

FIGURE 7.3 a. Ball and socket joint, shoulder

7.2.1.4 Gliding Joint (Figure 7.4)

The gliding joint is composed of small flat or concave/convex surfaces sliding over each other. The glide may be bi-directional and thus there are 2 degrees of freedom. The joints between carpal bones are gliding joints.

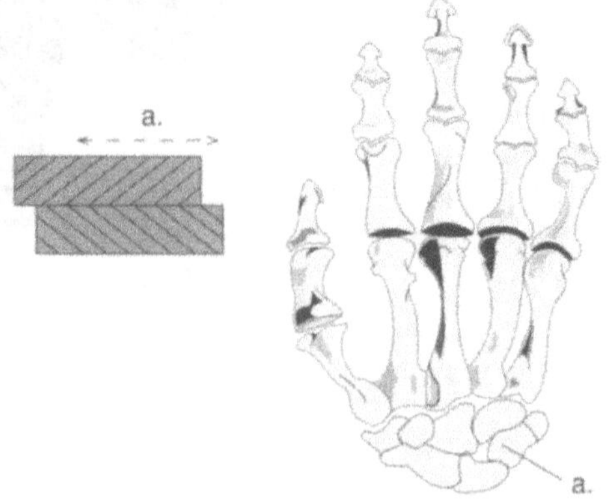

FIGURE 7.4 a. Gliding joint, joints between carpals

7.2.1.5 Condyloid Joint (Figure 7.5)

This joint, which is a unit fitting into an elliptical cavity, allows movement in two planes at right angles. There are 2 degrees of freedom. An example of a condyloid bone is the wrist.

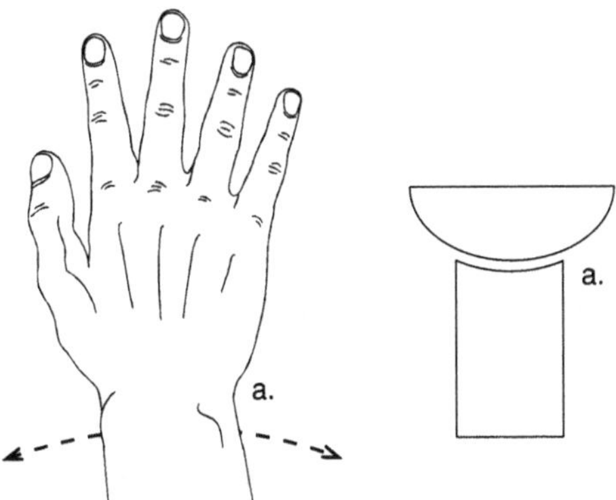

FIGURE 7.5 a. Condyloid joint, wrist

7.2.1.6 Saddle Joint (Figure 7.6)

This joint also displays movement in two planes at right angles and is composed of concave surfaces in contact. There are 2 degrees of freedom. The joint at the base of the thumb is a saddle joint.

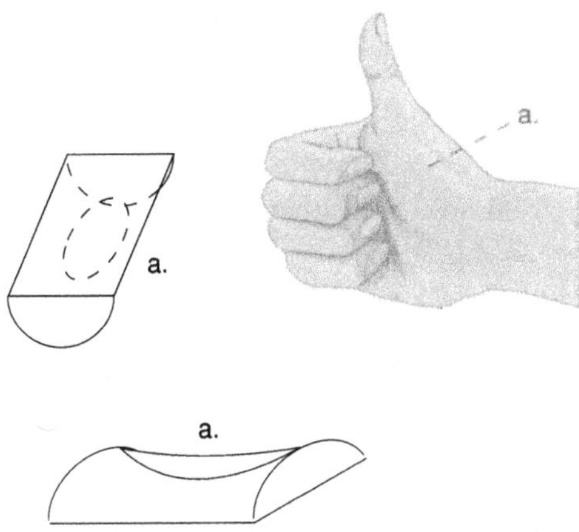

FIGURE 7.6 a. Saddle joint, base of thumb

7.2.2 Basic Human Movements

There are a number of basic human movements which apply to the entire anatomy and apply to the arm and hand as well.

7.2.2.1 Flexion

Flexion is a movement that decreases the angle between bones. Moving the calf towards the thigh is an example of flexion.

7.2.2.2 Extension

Extension is a movement which increases the angle between bones. Moving the calf away from the thigh is an example of extension.

7.2.2.3 Abduction

Abduction is a movement which moves a bone away from a midline. Moving the arm away from the torso, in the coronal plane, is an example of abduction.

7.2.2.4 Adduction

Adduction is a movement which moves a bone towards a midline. Moving an arm towards the torso, in the coronal plane, is an example of adduction.

7.2.2.5 Rotation

Rotation is a movement of a bone around an axis. Nodding the head in the "no" gesture is an example of rotation.

7.2.2.6 Circumduction

Circumduction is the movement of a bone so that the end of the bone traces a circle. Moving the arm to form a circle in the sagittal plane is circumduction.

7.2.2.7 Inversion

Inversion is moving of the sole of the foot inward towards the center line.

7.2.2.8 Eversion

Eversion is moving of the sole of the foot outward from the center line.

7.2.2.9 Protraction

Protraction is the moving of a part of the body forward along a plane parallel to the ground. Shrugging the shoulder forward is protraction.

7.2.2.10 Retraction

Retraction is the moving of a part of the body backward along the plane parallel to the ground. Shrugging the shoulder backward is retraction.

7.2.2.11 Elevation

Elevation is the raising of a part of the body. Shrugging the shoulder upward is elevation.

7.2.2.12 Depression

Depression is the lowering of a part of the body. Shrugging the shoulder downward is depression.

7.2.2.13 Pronation

Pronation is the rotation of the entire forearm so that the radius and ulna are crossed.

7.2.2.14 Supination

Supination is the rotation of the entire forearm so that the radius and ulna are parallel.

7.2.3 Movements of Arm and Hand

The following section describes and analyzes various motions of the upper extremity, in accordance with the principles discussed in the previous two sections and in accordance with anatomical structures described in Chapter 5.

7.2.3.1 Moving the Shoulder

The major function of the shoulder is support of and attachment for the arm. The shoulder does not possess extensive articulations, however, it displays four limited basic movements.

Elevation (Figure 7.7). The shoulder may be raised upward by muscular action upon the scapula. The muscles which elevate the shoulder are as follows: levator scapulae and trapezius.

Depression (Figure 7.7). The shoulder may be lowered by pulling downward upon the scapula. The muscles which depress the shoulder are as follows: pectoralis minor, pectoralis major, latissimus dorsi, and trapezius.

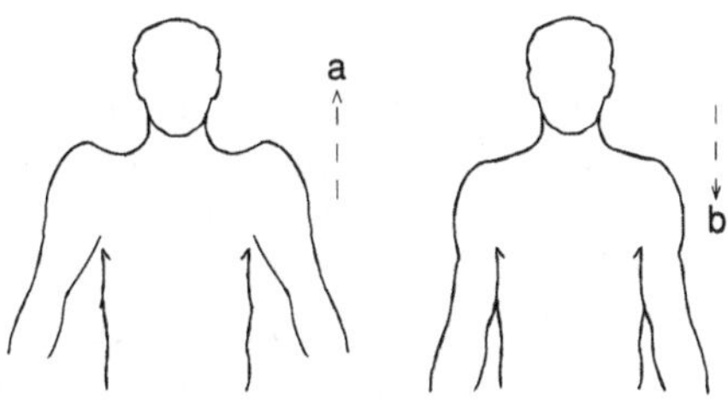

FIGURE 7.7

a. Shoulder elevation
b. Shoulder depression

Protraction (Figure 7.8). The shoulder may be moved forward by pulling at the scapula. The muscles which protract the shoulder are as follows: pectoralis minor and serratus anterior.

Retraction (Figure 7.8). The shoulder may be moved backward by pulling back on the scapula. The muscles which retract the scapula are as follows: rhomboid major and latissimus dorsi.

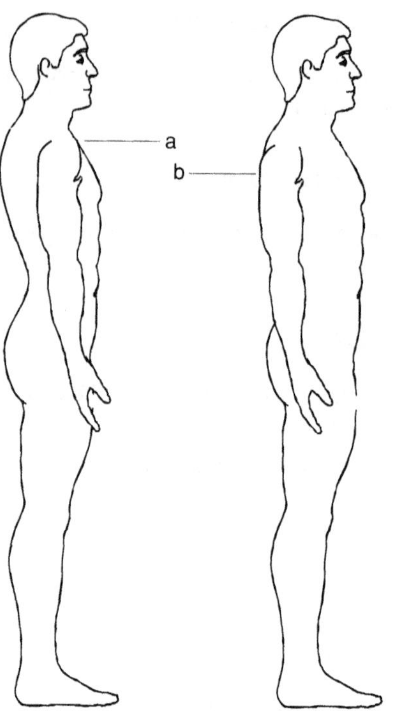

FIGURE 7.8

a. Shoulders protracted
b. Shoulders retracted

7.2.3.2 Moving the Upper Arm

The single bone (humerus) in the upper arm defines the grosser movements of the entire arm and provides support of the forearm. The head of the humerus forms a ball-and-socket joint with 3 degrees of freedom by fitting into the glenoid process of the scapula. There are five basic movements related to the upper arm.

Abduction (Figure 7.9). The arm in abduction is moved upward, away from the torso, in the coronal plane. Movement is the result of muscular action upon the upper part of the humerus. Muscles which abduct the arm are as follows: deltoid (middle) and suprasipnatus.

Adduction (Figure 7.9). The arm may be moved downward, towards the torso, in the coronal plane. Movement is the result of muscular action upon the upper part of the humerus. The muscle which adducts the humerus is as follows: pectoralis major.

Flexion (Figure 7.9). The arm may be moved upward, away from the torso, in the sagittal plane. Muscular action during flexion of the arm is provided by deltoid anterior and pectoralis major.

Extension (Figure 7.9). The arm may be moved downward, towards the torso, in the sagittal plane. Muscular action during extension of the arm is provided by deltoid posterior, teres major, triceps brachii, and long head.

Circumduction. The arm is capable of a gross circular movement which includes all aspects of the coronal and sagittal planes. This movement is activated by the muscles which were named above.

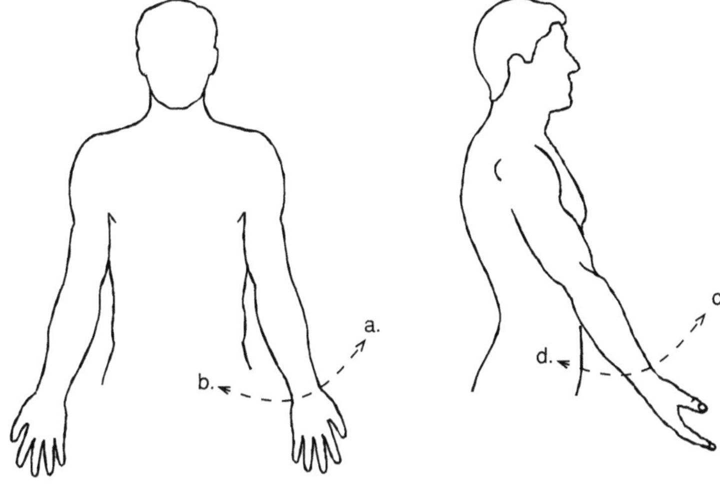

FIGURE 7.9

a. Abduction
b. Adduction
c. Flexion
d. Extension

7.2.3.3 Moving the Forearm

The forearm is an extension of the upper arm and a support for the hand. The forearm is composed of two bones: ulna and radius. The ulna has the primary function of supporting flexion and extension of forearm. The radius is involved with rotation of the forearm. The ulna forms a hinge joint, with 1 degree of freedom with the humerus. There are four movements related to the forearm.

Supination (Figure 7.10). The forearm is capable of longitudinal rotation. Supination is the movement in which the forearm rotates downward when the arm is extended in the sagittal or coronal plane. Supination is activated by the biceps brachii and the supinator located in the forearm.

Pronation (Figure 7.11). Pronation is the rotational movement opposite to supination. Pronation is motivated by pronator teres and pronator quadratus in the forearm.

Flexion (Figure 7.12). Flexion is the movement of the forearm towards the upper arm. Flexion is motivated by the biceps brachii, the brachialis, and the brachioradialis in the forearm.

Extension (Figure 7.12). Extension is the movement of the forearm away from the upper arm. Extension is motivated by the triceps brachii and the anconeus in the forearm.

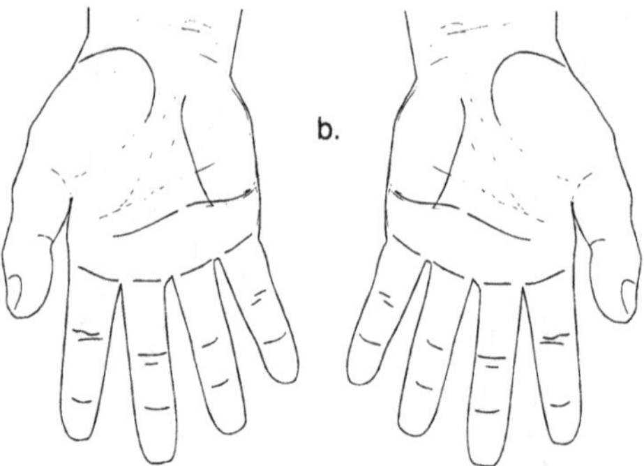

FIGURE 7.10 b. Hands supinated (smallest fingers together)

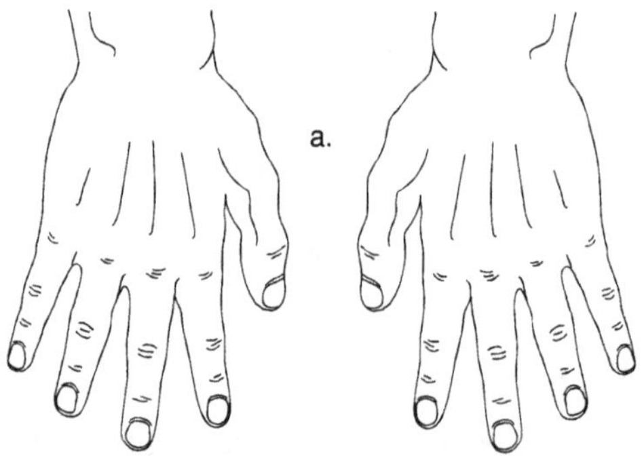

FIGURE 7.11 a. Hands pronated (thumbs together)

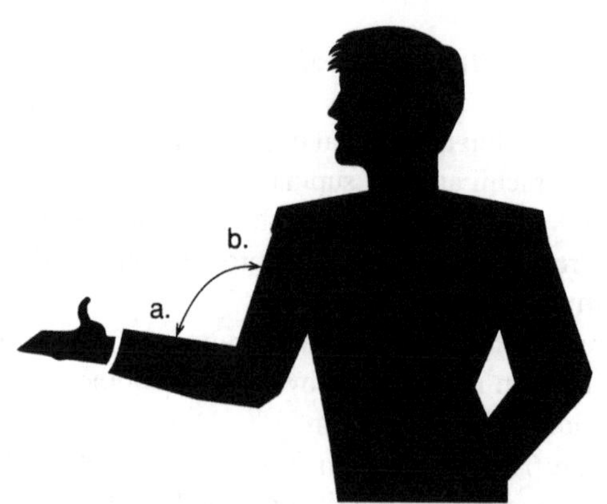

FIGURE 7.12 a. Extension of forearm
b. Flexion of forearm

7.2.3.4 Moving the Hand

The hand provides a surface against which the fingers exert pressure and, in turn, create a grasping force upon objects. The joint related to hand movement is the radio carpal joint. This joint is composed of the scaphoid, lunate, and triquetrum bones of the carpal tunnel and the radial styloid process at the distal end of the radius.

The joint is a condyloid joint with 2 degrees of freedom. There are five basic movements of the hand.

Palmar Flexion (Figure 7.13). Palmar flexion is the movement of the palm toward the soft aspect (belly) of the forearm. Palmar flexion is motivated by the flexor carpi radialis, flexor carpi ulnaris, and palmaris longus.

Dorsal Flexion (Figure 7.13). Dorsal flexion is the movement of the dorsal aspect (back) of the hand towards the dorsal aspect of the forearm. Muscles which move the hand in dorsal flexion are as follows: extensor carpi radialis brevis and extensor carpi ulnaris.

Radial deviation (Figure 7.14). Radial deviation is the movement of the hand towards the radial bone. This movement is activated by extensor carpi radialis brevis, extensor carpi radialis longus and abductor pollicis longus.

Ulnar deviation (Figure 7.14). Ulnar deviation is the movement of the hand towards the ulna bone. This movement is activated by extensor carpi ulnaris.

Circumduction. There is a limited amount of circumduction at the wrist. This circumduction is motivated by the muscles mentioned above and is supplemented by pronation/supination of the forearm.

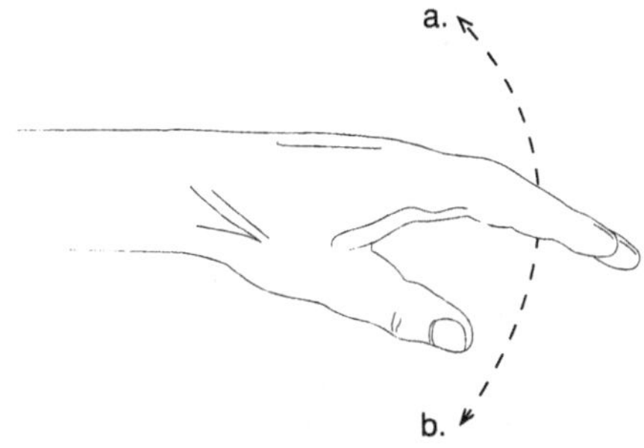

FIGURE 7.13

a. Dorsal flexion
b. Palmar flexion

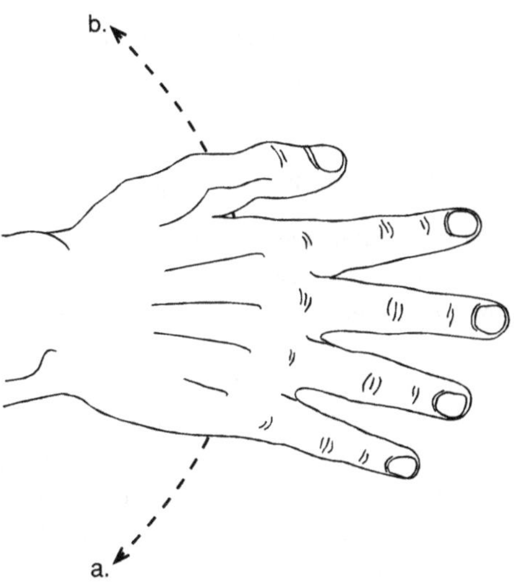

FIGURE 7.14

a. Ulnar deviation (towards smallest finger)
b. Radial deviation (towards thumb)

7.2.3.5 Movement of the Thumb

The movement of the human "opposable thumb" allows grasping of objects. The joint at the base of the thumb is a unique saddle joint with 2 degrees of freedom There are four basic movements of the thumb.

Flexion (Figure 7.15). Flexion of the thumb is the movement of the thumb away from the radius and towards the index finger. Flexion is activated by flexor pollicis longus within the forearm, and abductor pollicis brevis, opponeus pollicis, and flexor pollicis brevis within the palm.

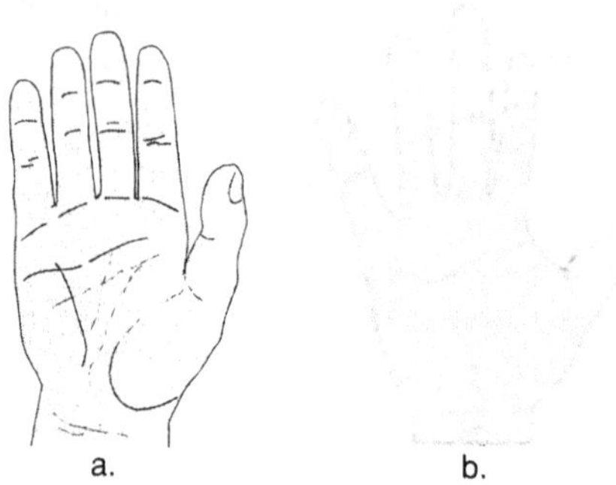

FIGURE 7.15

a. Thumb flexed and fingers adducted
b. Thumb extended and fingers abducted

Extension. Extension of the thumb is the movement of the thumb away from the index finger and toward the radius bone. Extension of the thumb is motivated by extensor pollicis longus and extensor pollicis brevis in the forearm.

Abduction. Abduction of the thumb is the movement of the thumb towards the palm. The movement is activated by abductor pollicis longus in the forearm and abductor pollicis brevis and opponeus pollicis in the hand.

Adduction. Adduction is the movement of the thumb away from the palm. This movement is activated by extensor pollicis longus in the forearm and adductor pollicis in the hand.

7.2.3.6 Movement of Finger

The joints (knuckles) at the base of the fingers are condyloid joints with 2 degrees of freedom. The joints between the individual segments of the fingers (phalanges) are hinge joints with 1 degree of freedom. Fingers possess four movements.

Flexion (Figure 7.16). Flexion is the movement of the fingers toward the palm. The muscles which motivate these movements are as follows: flexor digitorum superficialis in the forearm.

Extension (Figure 7.16). Extension is the movement of the fingers away from the palm. Muscles which motivate this movement are as follows: extensor digitorum, extensor digiti minimi, and extensor indicis in the forearm.

Abduction/Adduction (Figure 7.15). Abduction is the movement of spreading and separating the fingers. Adduction is the movement of bringing the fingers together. Interossei dorsalis muscles in the hand abduct fingers. Interossei palmaris muscles in the hand adduct fingers.

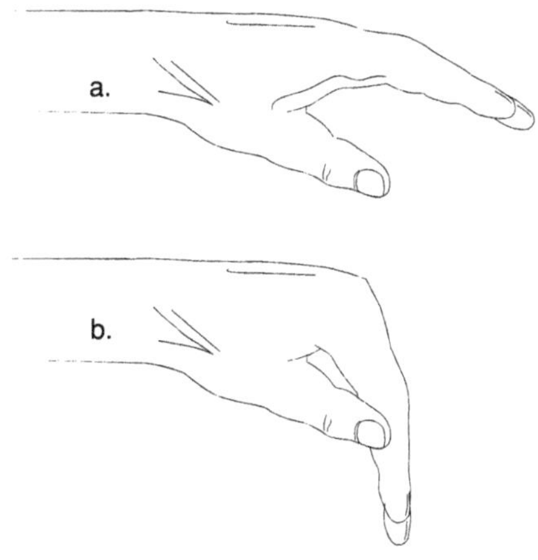

FIGURE 7.16 a. Fingers extended
b. Fingers flexed

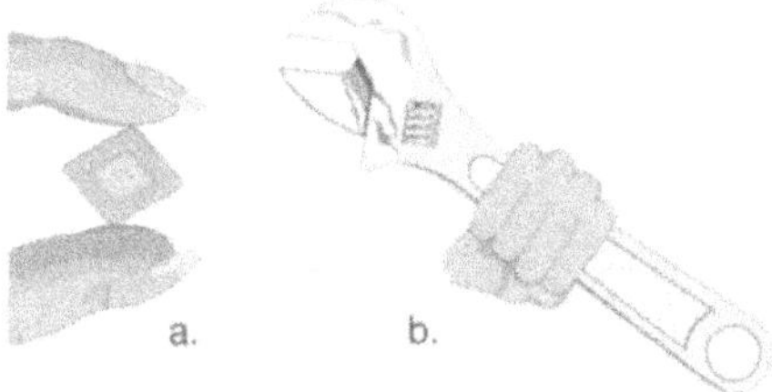

FIGURE 7.17 a. Pinch grip, object between tips of fingers
b. Power grip, object against palm and surrounded by fingers

7.2.4 Manipulation

Ultimately, the hand is used as a manipulative device for large and small objects, many of which are tools. In almost all cases the "opposable thumb," for which is credited a large part in the development of culture and civilization, will abduct towards palm or fingers, and fingers will flex towards palm and thumb. The final position is referred to as a grip or grasp. Kroemer, et al. (1990), indicated ten types of "couplings between hand and object." These types will include any possible posture the hand may assume towards an object (tool). These coupling are as follows: finger touch, palm touch, hook grip, tip grip, pinch grip, side pinch, writing grip, disk grip, enclosure, and power grasp. These couplings most related to large or small tools are as follows: pinch grip and power grasp (Figure 7.17).

8 Biomechanics and Hand Tools

8.1 INTRODUCTION

Biomechanics is the discipline which utilizes the principles of physics in order to explain the structures and functions of biological systems and organisms. (Özkaya and Nordin (1991) describe some taxonomies related to physics. They explain that one subdivision of physics, known as rigid body mechanics, deals with relatively undeformable objects in motion or at rest, being acted upon by internal or external forces.

Rigid body mechanics is further subdivided into statics, which is concerned with stationery objects, and dynamics which deals with moving objects. Dynamics is still further subdivided into kinematics and kinetics. Kinematics is concerned with movement only. Kinetics is concerned with movement as well as the forces related to the movement in question. Kinesiology is more concerned with kinematics and biomechanics is more concerned with kinetics. Biomechanics often involves quantifications related to forces.

8.2 BASIC MECHANICAL FUNCTIONS

There are several basic mechanical functions which are also used as basic elements in the configuration of animate and inanimate machines. The basic machine elements are considered to be levers (three categories), pulleys (two categories), inclined planes, and screws (Figures 8.1, 8.2, 8.3, 8.4). The pulley, and particularly the lever, are frequently discovered in musculoskeletal systems. These two simple forms of machinery will be separately discussed.

8.2.1 PULLEYS

The primary function of a pulley is to transmit force and to change its direction of movement within a machine. A pulley system is dependent upon three components: 1) the pulley, which is a wheel often with a continuous groove in its rim, 2) a rope, and 3) a source of energy which produces a pulling force. The rope is attached to the pulling force and also inserted around the pulley rim. The rope is also attached at its farthest end to an object which is to be moved. An adequate pulling force upon the rope will move the object in a direction other than the direction of the force. The pulley is, thus, an expedient device for changing the direction of an exerted force upon a remotely located object.

The aforementioned description was a description of a fixed pulley, that is, a pulley which is anchored to a non-moving surface. The other category of pulley is

FIGURE 8.1 Lever. Pry bar inserted into ground beneath log.

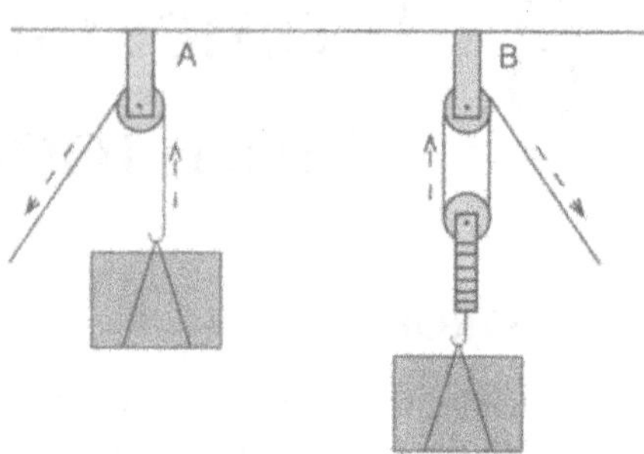

FIGURE 8.2 Pulleys.

A. Fixed pulley provides movement and change of direction but no mechanical advantage.
B. Moveable pulley also provides movement and change of direction as well as mechanical advantage.

FIGURE 8.3 Inclined Plane. Less force is needed to bring a load up to point A by using a ramp rather than by a direct lift. More distance, however, must be traversed by using the ramp.

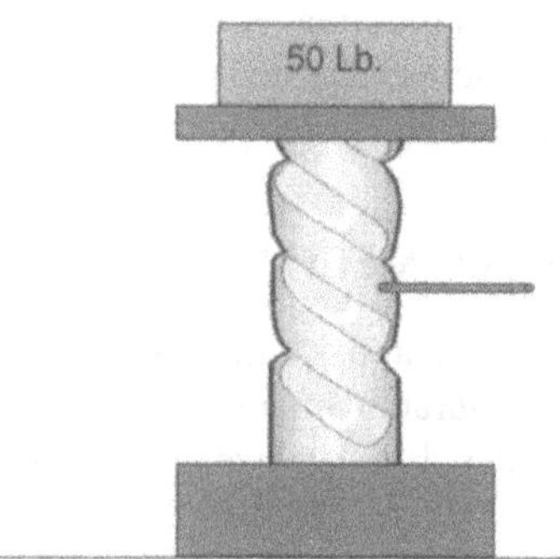

FIGURE 8.4 Screw. The screw is a variation of the inclined plane. After many turns of the screw the heavy object is gradually lifted.

known as the moveable pulley. This pulley is a system comprised of one or more pulleys which move along the rope. The moveable pulley, in addition to changing and transmitting force, is able to diminish the force needed to move the object. This ability to move an object with a force less than the weight of the object is referred to as mechanical advantage.

8.2.1.1 Pulleys within Biological Systems

As would be expected, there are literally no pulleys found in animal anatomies. There are, however, anatomical configurations in which force may be transmitted to a remote location with a resultant change in force. The human knee and human knuckle exemplify this anatomical design which approximates the functions of a pulley. Force is provided by contracting muscle. Tendons, which are extensions of the muscle and which are analogous to rope in a pulley system, pass over a smooth, often slotted bony surface which is analogous to a pulley. The contracting muscle pulls at the tendon which is attached to the tibia bone in the calf or the phalange bone of a finger. The direction of force on the bone may be as much as 90 degrees to the original force exerted by the muscle. The force brings about extension of the tibia towards the femur, or extension of the phalange towards the dorsal aspect of the hand. This form of pulley, without wheels, is sometimes referred to as a sliding surface (Figure 8.5). Frictional problems found in this system are dealt with by smooth bone surfaces and production of lubricating synovial fluids.

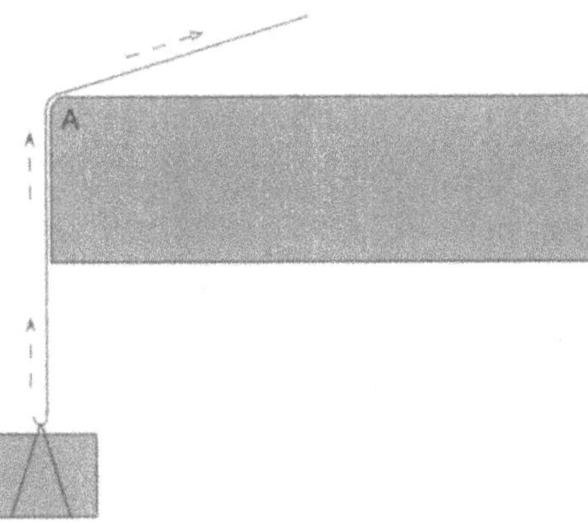

FIGURE 8.5 Sliding Surface. Rope (tendon) moves over a smooth rounded edge (A) to change direction of force. No mechanical advantage occurs.

8.2.2 Levers

There are three classes of levers. All three are included within the human anatomy and are used to transmit forces with mechanical advantage or disadvantage. There are four requisite components of a lever: 1) the lever arm which is a relatively long, rigid, unbreakable object, 2) a fulcrum which is a point on which the arm pivots, 3) a force placed upon any part of the arm except the fulcrum, and 4) an object which is to be moved. In a biological system, the lever arm is a bone, the fulcrum is a joint, usually at the end of the bone, the force is a contracting muscle, and the object being moved is a segment of the anatomy and/or object attached to the anatomy segment. The three types of levers are classified by the relative positions of fulcrum, force, and object being moved.

8.2.2.1 First Class Lever (Figure 8.6)

The first class lever has the following characteristics: 1) the fulcrum is located some–place on the arm but never at either end of the arm, and 2) the force being applied upon the arm and the object being moved are located at opposing sides of the fulcrum. A children's seesaw is an example of a first class lever within the anatomy. The head resting on the atlantooccipital joint, when nodding "yes," is a first class lever. The lever is usually deployed so that it provides a mechanical advantage in that the force needed to move the object is less than that of the object itself.

8.2.2.2 Second Class Lever (Figure 8.7)

The second class lever has the following characteristics: 1) the fulcrum is located at one of the ends of the arm, 2) the object being moved is located between the force placed upon the arm and the fulcrum. An example of a second class lever is

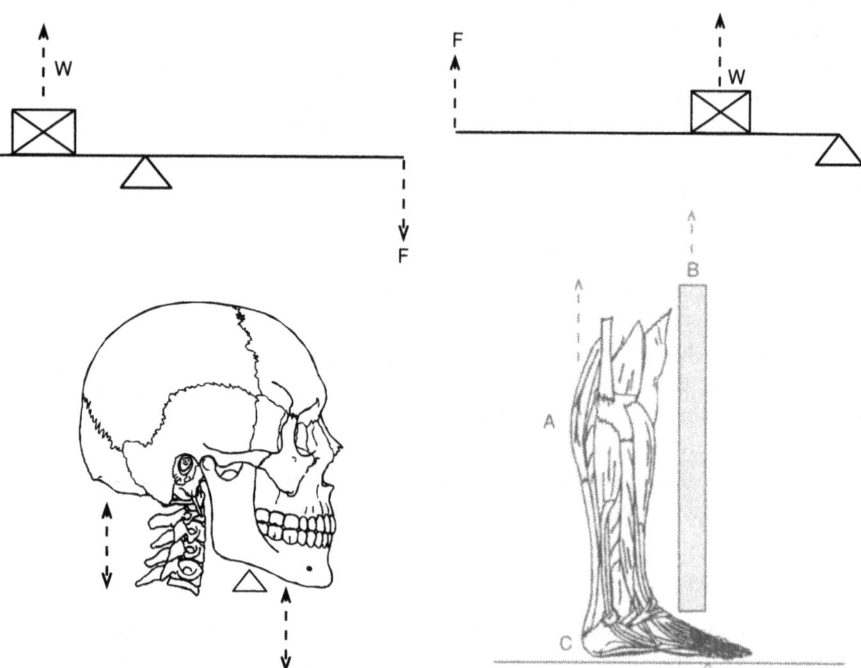

FIGURE 8.6 First Class Lever. Fulcrum located between ends. Force (F) is down and weight (W) moves up. The skull nodding "yes" is an example of first class lever.

FIGURE 8.7 Second Class Lever. The fulcrum is at one end, force (F) at the other end. Weight (W) is in between. The upward force moves the weight upward. The calf muscles (A) pulling up the heel bone (C) lifts tibia (B) as well as the entire anatomy.

a pry bar inserted into the ground beneath a log in order to move the log. An example of a second class lever within the anatomy is the extension of the foot by the gastrocnemius (calf) pulling upon the heel bone during walking or jumping. The second class lever is deployed in such a way that it provides mechanical advantage.

8.2.2.3 Third Class Lever (Figure 8.8)

The third class lever has the following characteristics: 1) the fulcrum is located at one of the ends of the arm, and 2) force placed upon the arm is located between the fulcrum and the object being moved. An example of a first class lever is a catapult which hurls a missile at high speed. An example of a third class lever within the anatomy is the brachialis muscle flexing the forearm towards the upper arm. The third class is usually deployed in such a way that there is no mechanical advantage provided; however, levers of this class provide for a substantial amount of speed and a long range of movement at the end of a body segment.

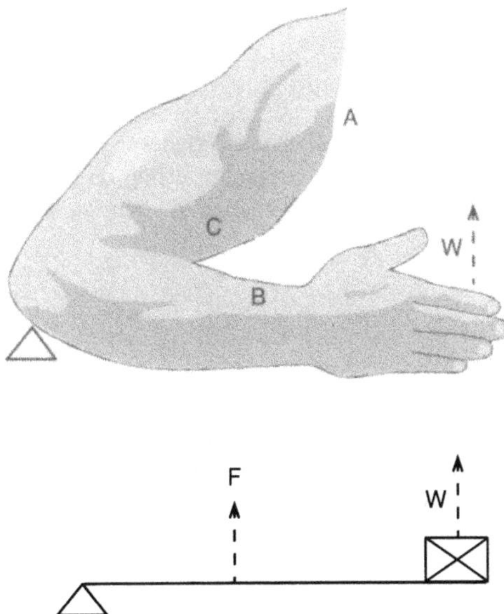

FIGURE 8.8 Third Class Lever. Fulcrum and weight (W) are at opposite ends. The force (F) is in between. The upward force moves weight up. The brachialis muscle (C) originating in humerus (A) and inserted into ulna (B) contracts and moves hand upward.

8.2.3 Levers in Arm and Hand

The third class lever is most frequently found in the human anatomy. This is also the case in the hand and in the arm. Some exceptions to this prevalence of third class levers are at the elbow joint where the triceps pull at the head of the ulna, thus, creating a first class lever, and at the hand where extensor and flexor muscles in the forearm, whose tendons are distally attached to finger tips, form second class levers.

8.2.4 Calculations

Calculations applied to levers are the same for all three classes. The basic formula for relating forces, weights, and their locations upon the lever is as follows:

R (RA) = F (FA)

In which R (resistance) is the weight of the object being moved or the weight of the body segment in question, RA (resistance arm) is the distance of the object to the fulcrum, F is the force placed upon the lever and FA (force arm) is the distance from the fulcrum where the force is applied.

Example #1. The entire length of a first class lever is 6 feet. There is a 100-lb weight resting on one end. There is a fulcrum located 2 feet from the weight. How much downward force is needed on the opposite end to move the 100-lb object?

R (RA) = F (FA)
100 (2) = F (4)
F = 50 pounds

Example #2. The entire length of a second class lever is 6 feet. There is a fulcrum located at one end. There is a 100-lb weight resting on the lever 2 feet from the fulcrum. How much upward force is needed at the opposite end to move the 100-lb weight?

R (RA) = F (FA)
100 (2) = F (6)
F = 33.3 pounds

Example #3. The entire length of a third class lever is 6 feet. There is a fulcrum at one end and a 100-lb weight at the other end. How much upward force must be applied at a distance 2 feet from the fulcrum to move the 100-lb weight?

R (RA) = F (FA)
100 (6) = F (2)
F = 300 pounds

8.3 SUMMARY

The above mentioned principles related to levers applies to physical, inanimate systems, and biological systems as well. Biomechanical issues generally related to human factors and ergonomics and specifically to hand tools are basically related to the following:

The physical juxtaposition of bone to muscle
The physical juxtaposition of bone to bone
The load (weight) placed upon the lever arm
The potential force which moves the lever arm

Among these four issues, the first, related to juxtaposition of muscle and bone, is beyond the control of industrial designers. The point at which muscle inserts into bone cannot be changed. The other three issues, however, are to some extent, under the control of industrial design. These three issues are translatable into four principles of design. These principles apply to all three lever systems within the anatomy including levers of the third class which predominate in their number.

1. Keep the resistance arm as short as possible. A short resistance arm requires less force and muscular effort upon the lever system and produces less fatigue, for the same weight, than a long "cantilevered" resistance arm (Figure 8.9).

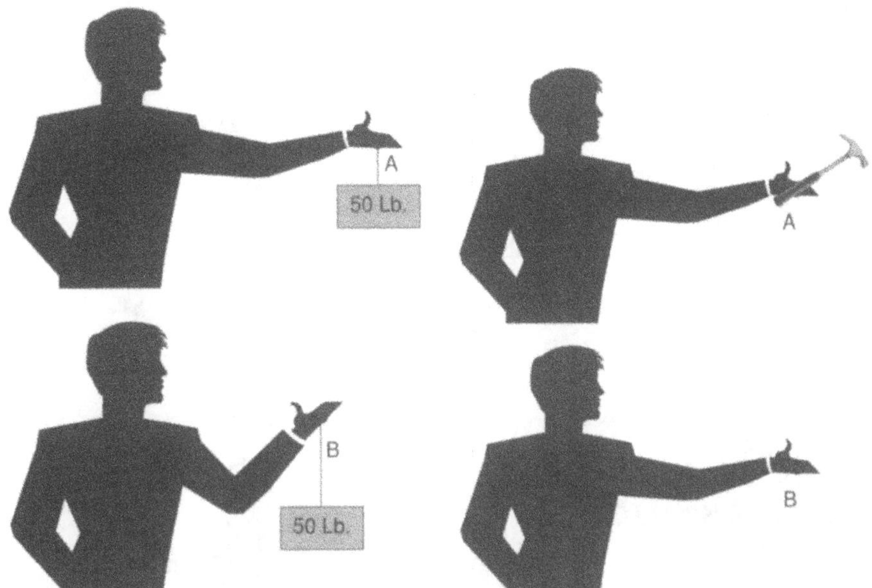

FIGURE 8.9 Decrease Resistance Arm. Suspend object as close to the fulcrum as possible.

A. Bad
B. Better

FIGURE 8.10 Resistance Arm Increased In Length By Tool. Force requirements are increased by extending a long object.

A. Bad
B. Better

2. Do not increase the resistance arm by coupling the resistance arm to excessively long objects. Increasing the resistance arm length with long objects increases the need for muscular effort upon the lever system and produces greater fatigue (Figure 8.10).
3. Do not couple the resistance arm with excessively heavy objects. Heavy objects, even on the end of short resistance arms, increase muscular effort and promote fatigue (Figure 8.11).
4. Do not design a lever system in which the bones form an excessively acute or excessively obtuse angle between themselves. A juxtaposition of bones in which the angle between bones is at the approximate midpoint of the overall articulation between the bones provides the best contractile potential and greatest strength for the muscle. In addition, an excessively acute articulation potentially places a dislocating hazard upon the joint (Figure 8.12).

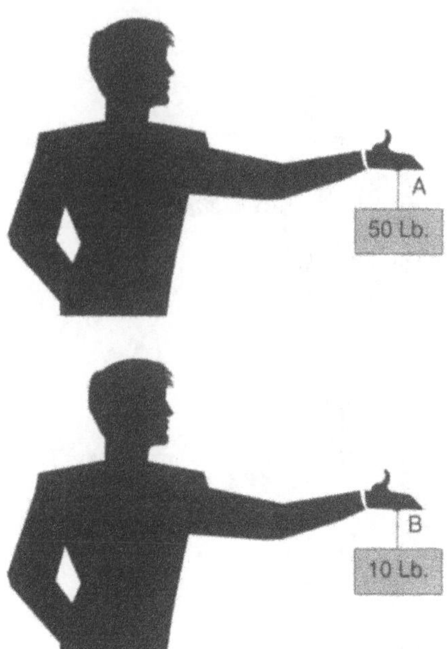

FIGURE 8.11 Excessive Weight Upon System. Diminish suspended weights.

A. Bad
B. Better

FIGURE 8.12 Angles of Articulation. Excessively acute or obtuse angles between bones are not as effective as an approximate 90° relationship.

A. Humerus	1. Bad (135°)
B. Ulna	2. Bad (45°)
	3. Better (90°)

9 Ergonomics and the Hand Tool

9.1 DISCUSSION

Definitions and descriptions of the term ergonomics vary from very broad to very narrow. Broader descriptions of ergonomics take consideration of the total human and include psychological/behavioral issues dealing with sensation, perception, cognition and decision-making in addition to questions related to the body which include issues dealing with anatomy, physiology, kinesiology and biomechanics. This chapter will deal with a narrower view of ergonomics because parts of the broader view are dealt with in other chapters, and because OSHA, as well as safety and health professionals, in general, are oriented towards the narrower view.

9.2 SCOPE OF ERGONOMICS

In consideration of the above, a specific definition of ergonomics is the discipline which studies how human beings physically posture themselves in relationship to the many different types of artifacts they have created. The artifacts vary in size and complexity and may include vehicles, buildings, furnishings, equipment, materials, controls, implements, and tools. The ergonomist applies the principles of anatomy, physiology, kinesiology and biomechanics to achieve a major mission: the control of musculoskeletal disease.

Additional missions include the control of other diseases such as cardiovascular disease, as well as, the goal of providing a comfortable environment which leads to efficiency and fewer errors and mishaps.

9.3 ERGONOMICS IN DESIGN

The principles of ergonomics, particularly those of its subdiscipline, biomechanics, are applied to the design of various human/artifact systems to control the incidence of musculoskeletal disease. These design activities pertain to the creation and use of hand tools as well as any other artifacts used by humans. Tools, when adhering to ergonomic design, particularly if their configuration departs from normal configuration, are frequently referred to as "ergonomic tools." The use of this term may be a misnomer. A tool, designed according to ergonomic principles, may appear to have a logical design which will diminish biomechanical stresses, however, the tool should not be referred to as an "ergonomic tool" unless follow up studies validate whether the tool has actually diminished the incidence of musculoskeletal disease. Only a small number of "ergonomic tools" have been validated (Tichauer 1978);

this small amount of validation does not, however, indicate that efforts at ergonomic designs in tools should be eliminated. Tools logically designed to conform to biomechanical principles adhere to a construct validity in which various universal theories and principles are applied to a design. Such logically designed tools display a substantial potential for disease control and may eventually be validated for their overall effectiveness. It is advisable to refer to the tools resulting from ergonomic design efforts not as ergonomic tools but as ergonomically-oriented tools.

9.4 ERGONOMICALLY-ORIENTED HAND TOOL DESIGN

Various professionals such as Chaffin and Andersson (1991), Alexander and Pulat (1985), Sanders and McCormick (1993), and Tichauer (1978) have outlined some basic principles of effective hand tool design. These principles, though few in number, are adequate for effective, healthful design. The following basic principles will be individually discussed: maintain a straight wrist, provide an optimal grip span, avoid tissue compression, and protect from heat, cold, and vibration.

9.4.1 Maintain a Straight Wrist

One of the most important requisites in tool design is keeping a straight alignment of the wrist while the tool is being used. The chapter on anatomy has indicated four basic movements at the wrist: palmar flexion, dorsal flexion, radial deviation, and ulnar deviation (Figure 7.10). Both palmar flexion (0 degrees to 90 degrees) and dorsal flexion (0 degrees to 99 degrees) indicate a high degree of articulation. Ulnar deviation (0 degrees to 47 degrees) displays moderate articulation and radial deviation (0 degrees to 27 degrees) only small articulation.

As described in the chapter on pathology, any of these movements, particularly those at the extremes in the range of movement, place extensive pressures upon the flexor tendons passing through the carpal tunnel causing inflammation of tendon sheath and ultimate pressure upon the median nerve. Ulnar deviation is most likely to be extensively assumed when using a tool such as a screwdriver or pliers in order to align the long axis of the tool with the long axis of the forearm. Several tool handle designs have been created which will align the long axis of the tool with the long axis of the forearm without requiring the hand to assume ulnar deviation. These tool designs are the slightly deviated head design, the slightly deviated handle design, the perpendicularly deviated head design, the perpendicularly deviated handle design (pistol grip), and the cylindrical tool design (Figures 9.1, 9.2, 9.3, 9.4, 9.5).

9.4.2 Provide an Optimal Grip Span

Mechanical disadvantages in grip force occur if the fingers are excessively flexed around a small diameter handle or minimally flexed around a large diameter handle. These mechanical disadvantages lead to excessive grip force requirements which in turn lead to fatigue tendinitis and the hazard of accidentally dropping the tool (Figure 9.6).

FIGURE 9.1 Bent Nose Pliers. Such tools permit grasping, cutting, or turning objects while the wrist remains in a relatively straight position.

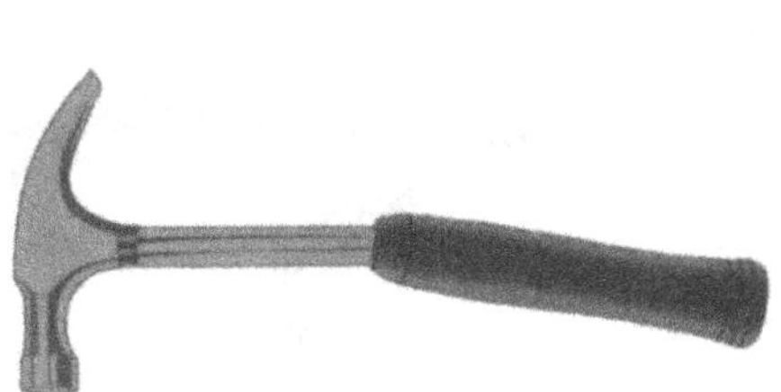

FIGURE 9.2 Hammer With Deviated Handle. The slightly bent handle maintains a straight wrist during the final impact position.

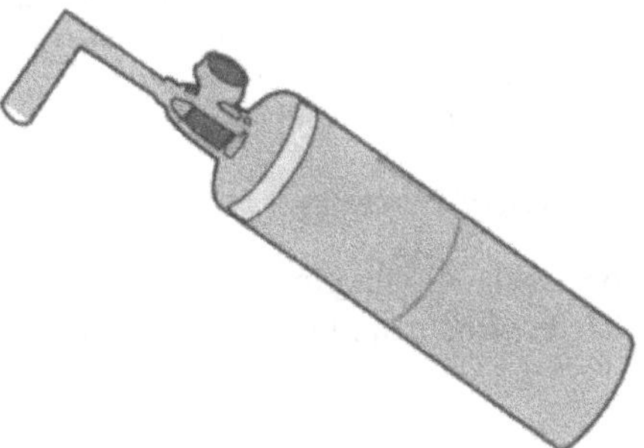

FIGURE 9.3 Soldering Iron. The perpendicular bend of the head permits application of heat to a distal object while avoiding deviation of the wrist.

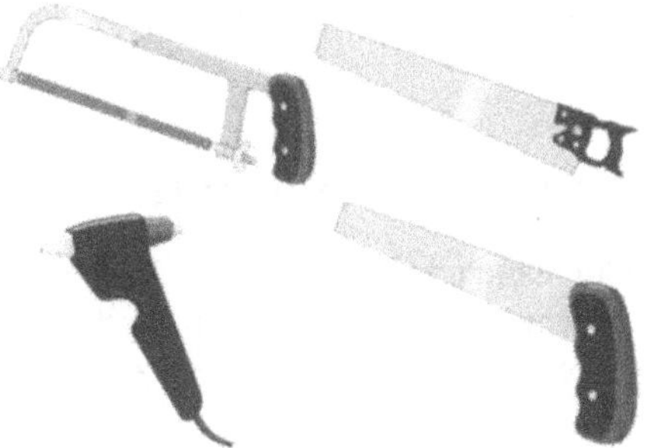

FIGURE 9.4 Saws, Power Wrench, Knife. A perpendicular handle (pistol grip) maintains a straight wrist during cutting, sawing, or rotary operations such as drilling or nut tightening.

FIGURE 9.5 Cylindrical Handle. A tool for rotary action on a horizontal work piece maintains the wrist in a straight position.

9.4.3 Avoid Tissue Compression

Gripping forces placed upon handles which are short and possess small surface areas cause ischemia (loss of local circulation) in palmar tissues. Handles of adequate length and surface area, covered with compressible material such as plastic or rubber, should be used to prevent loss of circulation (Figure 9.7).

FIGURE 9.6 Excessive Grip Span. Excessive grip span does not allow optimal application of force and imposes undue stress upon the joints.

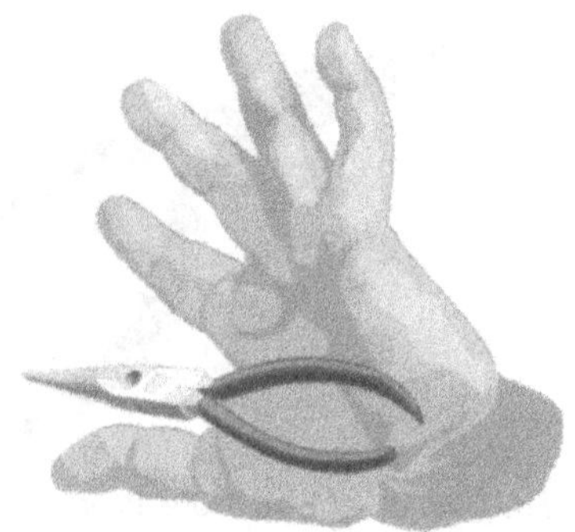

FIGURE 9.7 Small Surface Area. Excessively thin or short handles cause small surface areas which in turn cause excessive pressure on tissues, leading to loss of local circulation.

9.4.4 Protect Against Heat, Cold, Vibration Extremes

Circulation is also affected by other factors such as vibration and temperature extremes from the tool or from the immediate environment. Gloves and insulating materials are used as protective devices.

9.5 SUGGESTED DIMENSIONS AND CONFIGURATIONS

Table 9.1 contains a compilation of dimensions and configurations from a large number of sources; Eastman Kodak Co. (1983), Eastman Kodak Co. (1986), Greenberg and Chaffin (1976), Mital and Kilbom (1992), and Robinson and Lyon (1994). This information covers the majority of design questions.

9.6 CONSIDERING THE SHOULDER

The major thrust of tool design is the protection of wrist and hand. The use of the pistol grip design and the cylindrical tool design will maintain a straight wrist, but if used under incorrect circumstances will cause an undesired abduction of the arm. The correct choice of pistol grip or cylinder grip will keep the arm desirably adducted. The incorrect choice of pistol grip or cylinder grip will keep the arm undesirably abducted and may also deviate the wrist. The determination of tool grip is predicated upon the alignment of the workpiece (whether vertical or horizontal) and the elevation of the workpiece in relationship to the torso. Table 9.2 describes the correct choice of tool handle in relation to work piece.

9.7 OTHER CONSIDERATIONS

Other design features which may aid ergonomic design and contribute to safety are listed below.

TABLE 9.1
Suggested Dimensions and Configurations (Figures 9.8A-G)

Feature	Specification	Detail
Handle Configuration	Round or Oval	A
Handle Diameter for Power Grip	1.25"–1.75" Diameter	A
Handle Diameter for Precision Grip	.3"–.6" Diameter	A
Handle Length	4" minimum	A
Handle Length with Gloves	4.5" minimum	A
Pistol Grip Handle Angle	80°	B
Cylinder Grip Handle Angle	90°	C
Hammer Handle Off Set	10°	D
Grip Span Pliers, Cutters	2.5"–3.5"	E
Grip Curvature Pliers, Cutters	.5"	E
Bent Nose Pliers Offset	30°–40°	F
Unsuspended Power Tool Weight	5 pounds	—
Unsuspended Precision Tool Weight	4 Pounds	—
Trigger Span (use 2 or more fingers)	3.5" maximum	G
Handle Materials	Nonconductive, nonporous, non slip, smooth, slightly compressible	—
Handle Fluting, Screwdriver	Only if high torque needed	—
Secondary Handle	Only if torque resistance needed or excessively heavy tool	—

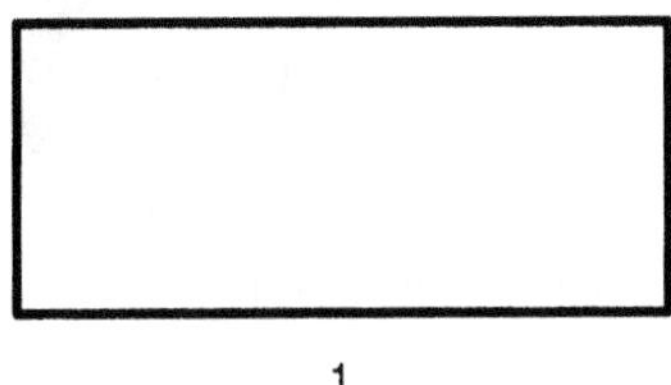

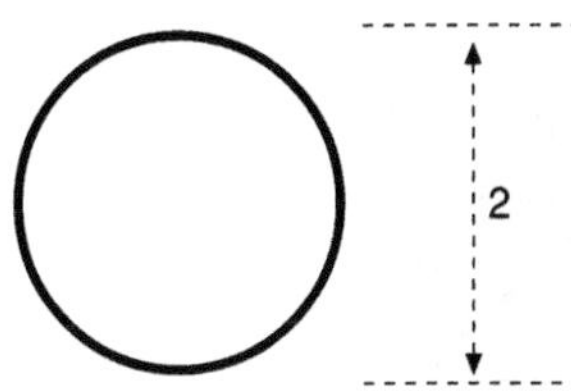

FIGURE 9.8 Detail A.

1. Handle length.
2. Handle diameter.

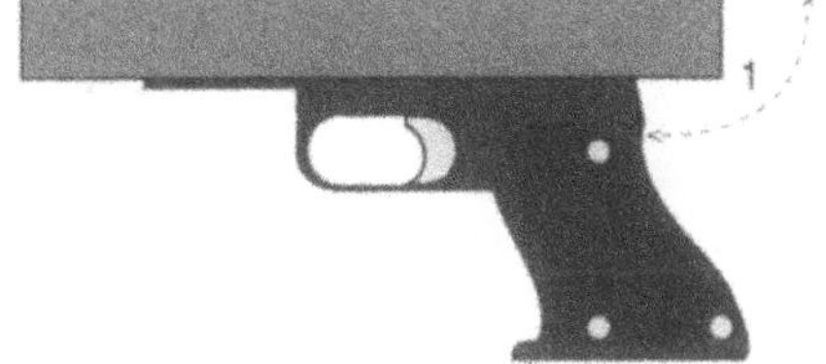

FIGURE 9.8 Detail B.

1. Pistol grip angle.

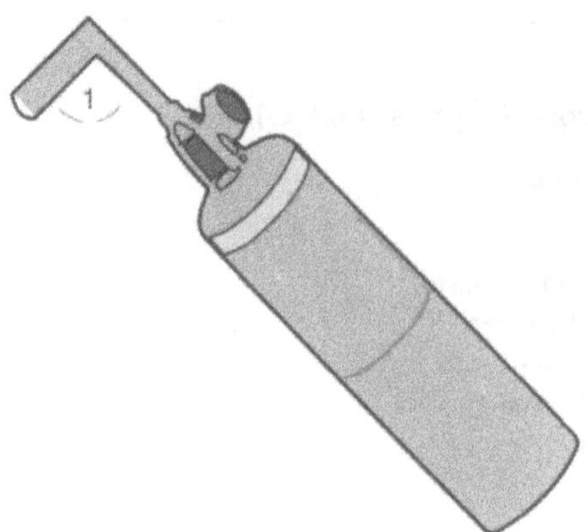

FIGURE 9.8 Detail C.

1. Working head angle for cylinder handle tool.

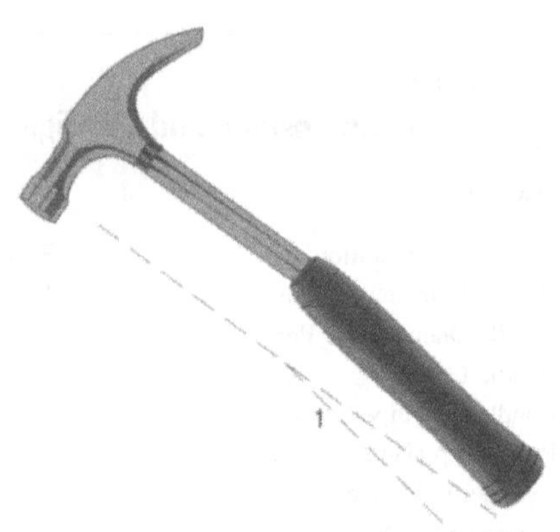

FIGURE 9.8 Detail D.

1. Bent hammer handle angle.

FIGURE 9.8 Detail E.

1. Handle span.
2. Handle curvature.

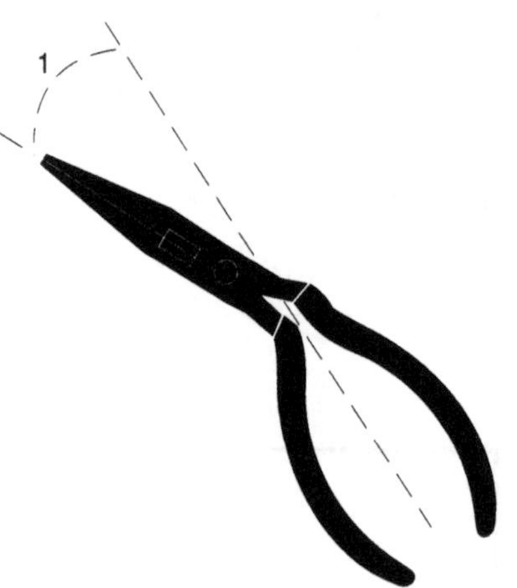

FIGURE 9.8 Detail F.

1. Bent nose angle.

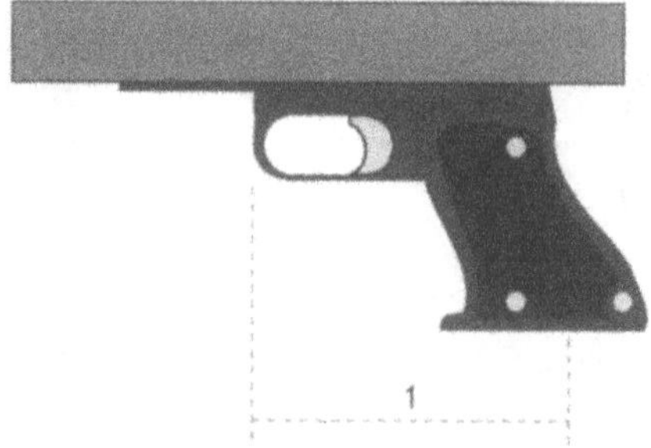

FIGURE 9.8 Detail G.

1. Trigger span.

TABLE 9.2
Choice of Tool Handle. Electric Drill

Workpiece Height	Vertical Workpiece	Horizontal Workpiece
Above Head	Cylinder Grip	—
Shoulder Level	Pistol Grip	Cylinder Grip
Waist Level	Pistol Grip	Cylinder Grip
Crotch Level	Cylinder Grip	Pistol Grip

9.7.1 Second Handle

An additional handle for powered tools, located near the front end, helps support a heavy tool, resistance to excessive torque, and safe placement of the tool on to the work piece.

9.7.2 Expanding Spring

A spring which automatically separates the handles of a two-handled tool will prevent the constant need of opening the handles (Figure 9.9).

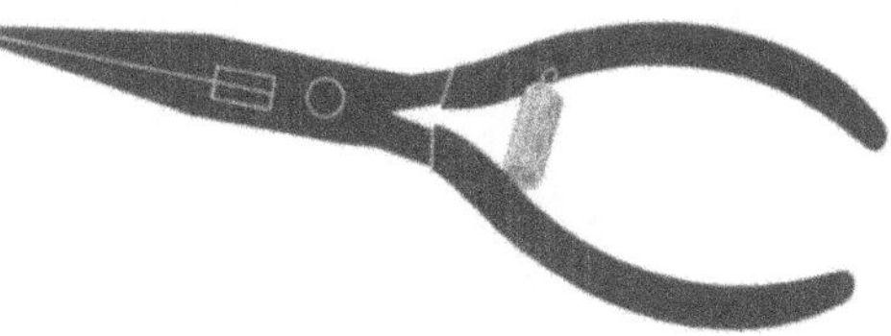

FIGURE 9.9 Expanding Spring. Expanding spring on handles opens handles without stressful operator effort.

9.7.3 Thumb Stop

A thumb stop at the top of one handle of a two-handled tool prevents slippage and provides added strength to tool usage (Figure 9.10).

9.7.4 Gloves

Gloves may be unavoidably used in inclement circumstances. Compensations must be made, when gloves are used, for a diminished grip force, diminished manual dexterity, and increased bulkiness which requires large handle dimensions.

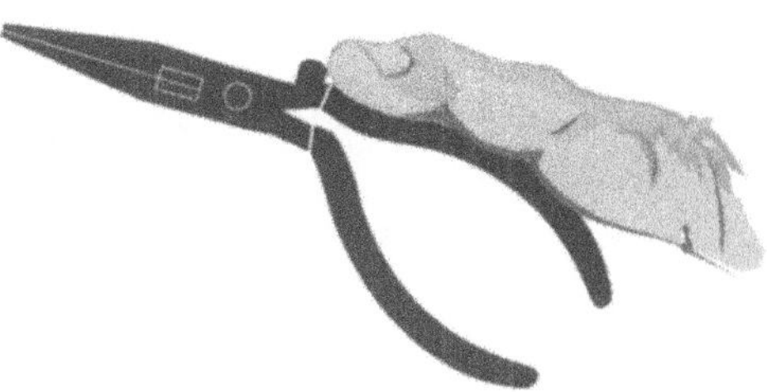

FIGURE 9.10 Thumb Stop. A thumb stop improves stability, increases thrust force, and avoids slippage.

10 Safe Design and Safe Use of Hand Tools

10.1 DISCUSSION

Trauma occurring to workers or consumers while using hand tools may be subdivided into two categories: instant (acute) trauma and cumulative (subacute) trauma. Cumulative trauma, which deals with ergonomic and biomechanical considerations of hand and tool, is covered in other chapters of this book. Instant trauma and the hand tool will be discussed in this chapter.

10.2 GENERAL INCIDENCE

Chapter 3 of this book, which deals with epidemiology, reports on the incidence of instant trauma as related to hand tools. The general conclusion is that the hand tool is not a frequent contributor to instant trauma, however, severity of injuries from hand tools are high and may lead to extensive disabilities and high worker compensation costs.

10.3 DEFINITION AND CATEGORIES

Instant trauma is an immediate mechanical disruption of tissues usually due to a collision of the tissue with a hard object. Instant trauma has been traditionally categorized as the following:

Abrasion: Scraping of tissue.
Laceration: Tearing of tissue.
Contusion: Impacting of tissue.
Incision: Cutting of tissue.
Puncture: Piercing of tissue.
Fracture: Breaking rigid tissue (bone).
Sprains/strains: Stretching and misalignment of tissue.
Burns: Destruction of tissue by excess heat.

10.3.1 Specific Incidence

Mital and Kilbom (1992) indicate that the preponderance of instant trauma is attributable to non-powered hand tools, particularly knife, hammer, wrench, and shovel, causing cuts lacerations, sprains and strains, to upper extremities, especially to fingers. Powered hand tools will less frequently cause instant trauma. Those powered hand tools most frequently associated with injury are saws, drills, grinders, and hammers. These powered tools most frequently cause cuts, lacerations, sprains, and strains.

10.4 GENERAL CONSIDERATIONS

There are a number of considerations related to safety and the hand tool. One important aspect is the functional assignment of the hand(s). Hands, when used with a tool, may be classified into the holding hand and the assisting hand (Figure 10.1).

FIGURE 10.1 Hammering Nail. Holding hand: hammer. Assisting hand: nail.

The holding hand grasps the tool, usually at the specially-designed handle and controls the movement or the stationary stability of the tool. The assisting hand may hold a supplemental handle on the tool (example: a large electric drill) and thus, provide added tool stability. The assisting hand may hold another tool such as a chisel. The assisting hand may also hold the work piece; thus, aiding tool control, or the assisting hand may act as a leaning support for the user's torso, thus aiding in tool control by providing overall balance for the user. In some cases there are two holding hands as in the case of tools with two major handles (scythe, pruning, and shears) or in the case of using two tools simultaneously (hammer and chisel). Generally, the safety of the holding hand depends upon the safe design of the hand tool, particularly the handle. Generally the safety of the assisting hand depends upon safe work practices of the user. In addition to the hands, other considerations of hand tool safety deal with the remainder of the user's anatomy: eyes, arms, legs, feet, and torso. These areas may also be vulnerable due to unsafe hand tool procedures. The following selections will individually deal with safe hand tool design and safe hand tool procedures.

10.5 SAFE DESIGN

There are several principles in tool design which may reduce cutting, laceration and contusion hazards during use of a tool. These principles should be observed when tools are designed and manufactured. These principles should be sought after by those who purchase tools for themselves or for their organization.

10.5.1 Loss of Grip

A major concern in safe tool design is loss of tool control by tool slippage from the holding hand. Slippage from the holding hand allows: 1) the tool to be dropped or thrown outward, causing injury to the user, other individuals, or the environment (Figure 10.2), and 2) a sliding movement of hand along the handle or other handhold on to a hazardous part of the tool such as knife blade or rotating drill bit (Figure 10.3). Some guidelines for handle or handhold design follow.

Handles and handholds should be of a circumference (neither too thin nor too thick) which allows the user to place optimal gripping force upon the handle. See the chapter on ergonomics for optimal dimensions.

For purposes of optimal grasp, handles, or handholds should be of a length which is slightly longer than the width of the palm of the holding hand. See the chapter on ergonomics for optimal dimension.

Prevention of slippage upon the surface of the handle or handhold necessitates a low coefficient of friction surface material such as rubber or foamed plastic. Slight compressibility of the handle material may also prevent slippage. Knurling, i.e., depressed contours which fit the fingers, will also avoid slippage provided the knurls are accurately fitted to each individual user's fingers. If this cannot be done and the knurls are not fitted exactly to the individual fingers, then ergonomic hazards will arise (see chapter on ergonomics).

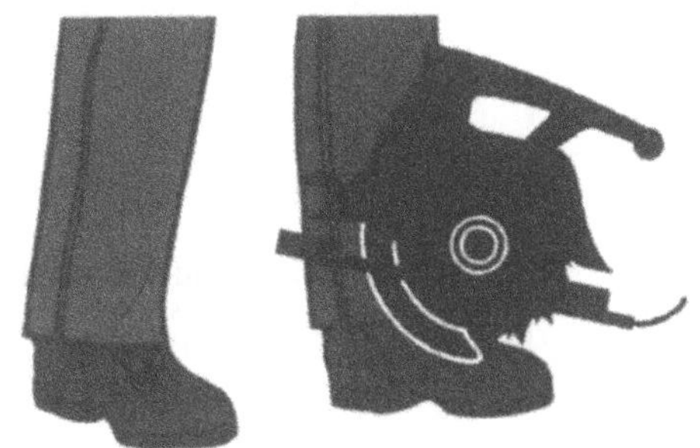

FIGURE 10.2 Dropped Tool. Power saw dropped on foot.

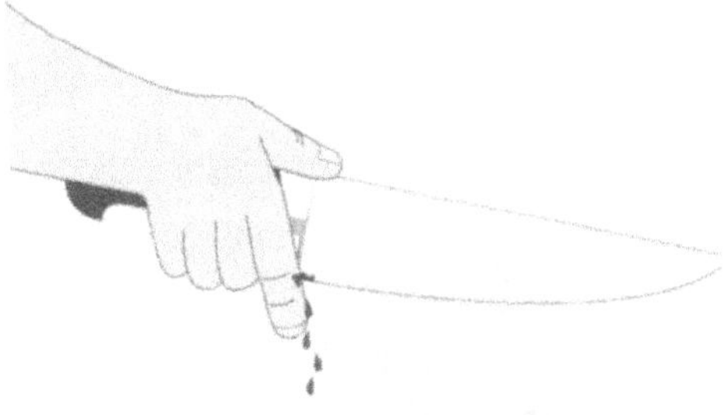

FIGURE 10.3 Slipping Hazard. Hand slips on to unguarded surface.

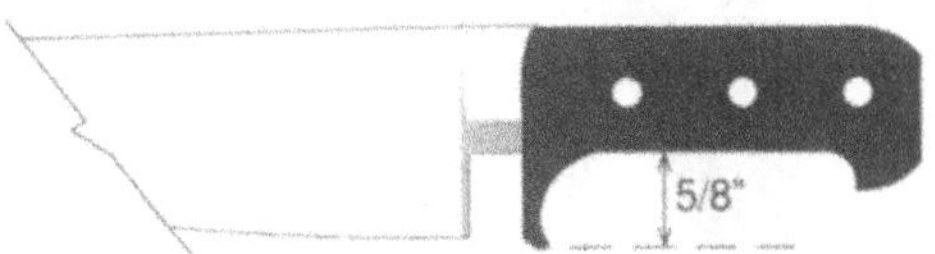

FIGURE 10.4 Curved Knife Guard.

To prevent slippage, place an adequately wide flange between the handle or handhold and hazardous working parts of the tool such as blades or moving parts. The flange in the immediate area of the index finger should project at least five-eighths inches out from the handle (Figure 10.4). Curving the flange downward towards the index finger will also provide additional protection from slippage.

Placement of an enclosure around the handle or handhold will prevent the holding hand from slipping on to the hazardous part of the tool. The enclosure may cover the entire hand or may take the form of a ring into which the index finger is inserted (Figures 10.5, 10.6).

Tools heavier than 25 pounds may readily slip from the grip. Large heavy tools should be designed so that their center of gravity is located in the same approximate area that they are to be grasped (Figure 10.7). Large heavy tools should be provided with an additional handhold for the assisting hand (Figure 10.8).

FIGURE 10.5 Knife. Ring guard. Enclosure for one finger.

FIGURE 10.6 Knife. Handle guard. Enclosure for all fingers.

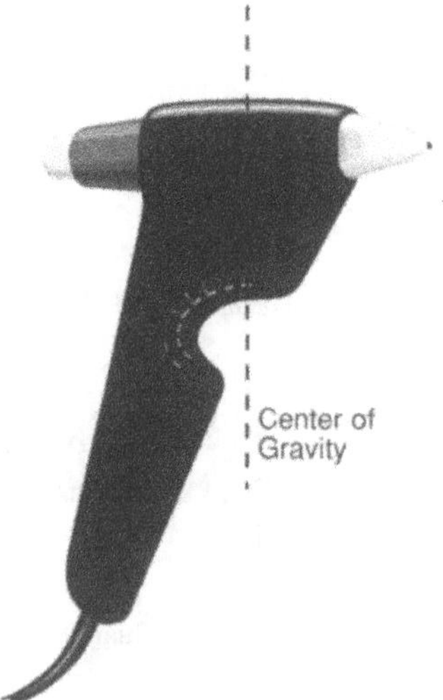

FIGURE 10.7 Balanced Power Tool. Handle located at center of gravity.

FIGURE 10.8 Second Handle. A second handle near the front of a large tool allows added control. A second handle will also aid in resisting high torque forces.

10.5.2 Crushing in Between

An additional hazard is crushing of the holding hand, or fingers between the handle and another handle or another object. The enclosure mentioned above, which prevents hand slippage, will also protect from crushing and pinching.

> An additional device in the form of a handle stop will also avoid crushing. A handle stop which prevents two handles from coming together to form a pinch point (as in the case of metal cutters) should project at least 1 inch from the handle (Figure 10.9).

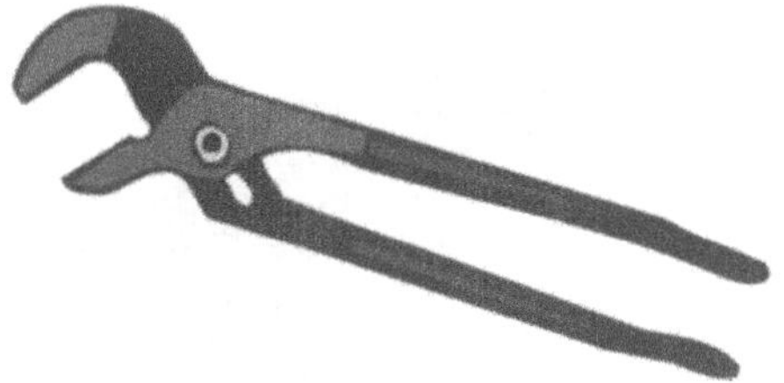

FIGURE 10.9 Handle Stops. Handle stops allowing at least 1 inch clearance prevent crushing fingers between handles.

10.5.3 Accidental Activation

Unexpected activation of a power tool may lead to a severe injury of the hand(s) if resting upon a moving part of the tool. There are several safeguards related to inadvertently activating mechanisms of power tools.

A "dead man's" trigger type switch, wide enough to be activated by only one finger, and which springs back into the "open/off" position, is the most desirable design. Clicking or sliding on/off buttons and toggle switches should be avoided.

A trigger guard around the trigger which allows only one finger to enter the guard prevents further accidental activation (Figure 10.10).

A foot switch, rather than a trigger, may be an effective device which prevents unexpected activation by the hand. A cover, to prevent inadvertent activation by the foot, should surround the foot switch at top and two sides (Figure 10.11).

A trigger activating button is an additional level of safeguarding. This form of safeguard will not allow a trigger to be operable unless a separate activating button is pressed.

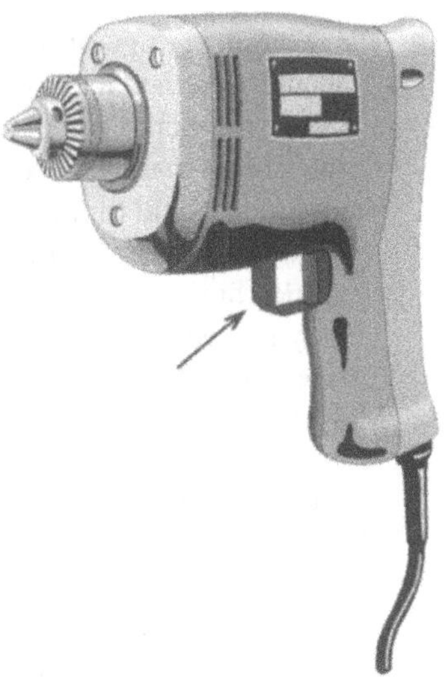

FIGURE 10.10 Trigger Guard. A trigger guard prevents accidental activation of a power tool.

10.5.4 Other Hazards

Other miscellaneous hazards may be controlled by the following designs.

Electrically powered hand tools should be grounded or double-insultated to prevent electric shock.

The overall configuration of a powered tool should be such that placing the tool on to a flat or projecting surface will not activate the tool. This hazard may also be avoided by providing eye hooks so that the tool may be hung up rather than laid down.

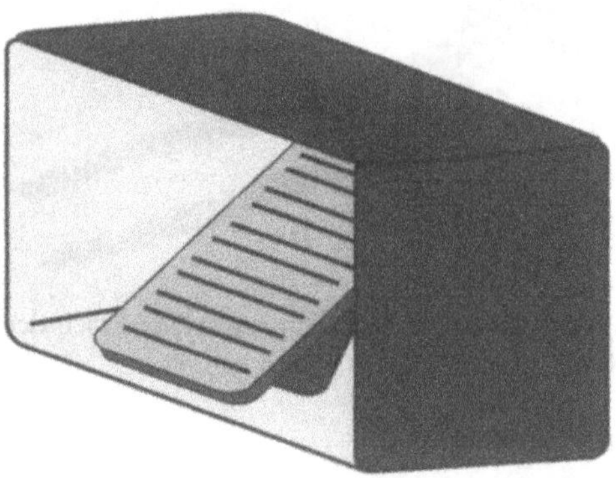

FIGURE 10.11 Guarded Foot Switch. Metal enclosure prevents inadvertent activation of the pedal switch.

The overall configuration of a tool should be such that the user can easily see the workpiece and thus have safe control over the operation.

Hand tools should be made of materials which produce minimal sparks during operation, particularly if used in a potentially explosive atmosphere.

Moving parts of powered tools, which may entangle clothing or hair, should be covered.

Moving working parts such as circular saw blades should be at least partially covered (guarded) to prevent contact with the assisting hand. The guard should be interlocked into the power source to prevent unauthorized removal of the guard (Figure 10.12).

In order to prevent cuts, sharp edges, particularly of square handles, should be rounded into at least a one quarter inch radius (Figure 10.13).

To avoid splitting or breaking, sound, hard wood such as ash, hickory, or maple should be used for hammer handles.

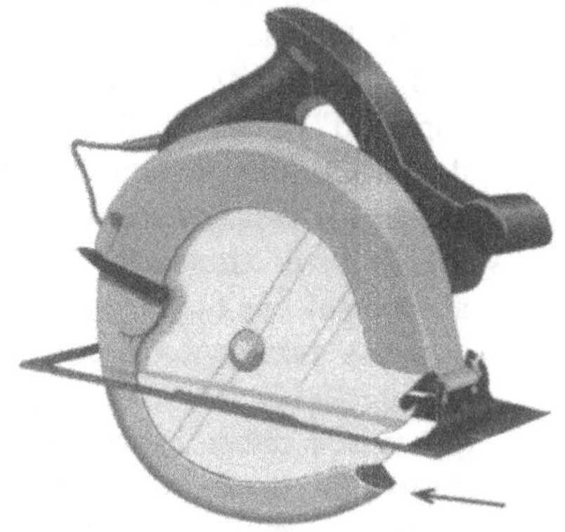

FIGURE 10.12 Electric Circular Saw. Exposed areas of blade covered by guard.

FIGURE 10.13 Rounded Edges. The edges of square or rectangluar profiled handles should have a one quarter inch radius to prevent excess pressure upon tissues.

10.6 SAFE USE

Safe use of powered and unpowered hand tools involves intelligent foresight on the part of the user. Using well maintained tools and using correctly designed tools for a particular job may prevent accidental injury. Some safe tool practices follow:

Eye protection, such as safety glasses, goggles, or face shields, should be used when tool use may produce flying particles (Figure 10.14).

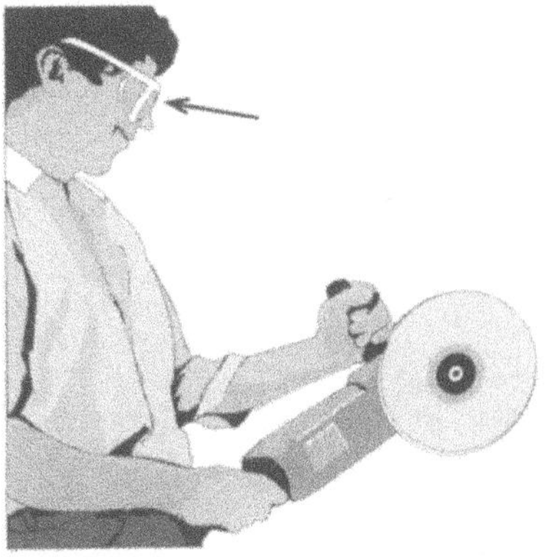

FIGURE 10.14 Eye Protection. In chipping and striking operations, safety glasses should at least be used. In some cases, goggles or a face shield may be even more appropriate.

Cutting tool usage should be such that the user will avoid moving the working part of the tool towards his/her torso or appendages. Some examples are pulling a knife inward towards the user's torso, pushing a powered circular saw blade on to the assisting hand, or pushing a chain saw downward towards the user's legs (Figure 10.15).

FIGURE 10.15 Cutting. Sharp cutting tools should be moved away from body parts.

Chisels should not be allowed to develop "mushroom heads" which may produce flying particles when struck (Figure 10.16).

Chisels should be of sufficient length so that the fingers of the assisting (holding) hand are not near the top of the chisel (Figure 10.17). Better still, a foam rubber guard should be placed at the top of the chisel or a separate holding device for the chisel should be used in order to prevent striking of the assisting hand by the hammer.

Wrenches of correct jaw dimensions which snugly fit the nut should be chosen in order to prevent accidental slippage and crushing of fingers between the handle and any hard object. It is usually desirable to pull a wrench towards the body rather than push a wrench away from the body towards an external hard object (Figure 10.18).

A "cheater bar" which increases the length of a wrench handle should not be used. Such a bar will increase forces which will cause slippage or a broken wrench handle (Figure 10.19).

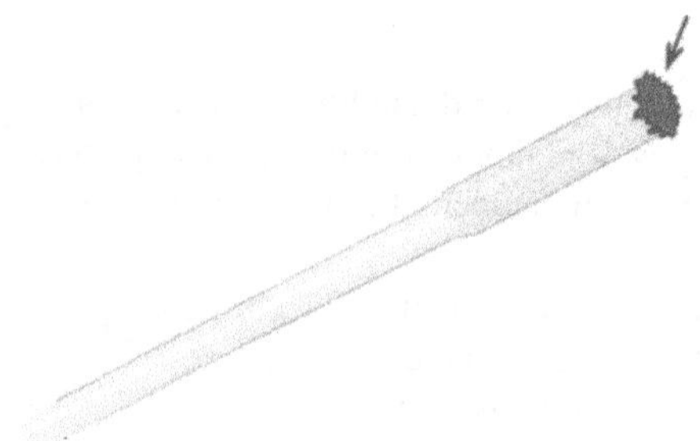

FIGURE 10.16 Mushroomed Chisel Head. Striking splayed, mushroomed chisel heads produces flying chips.

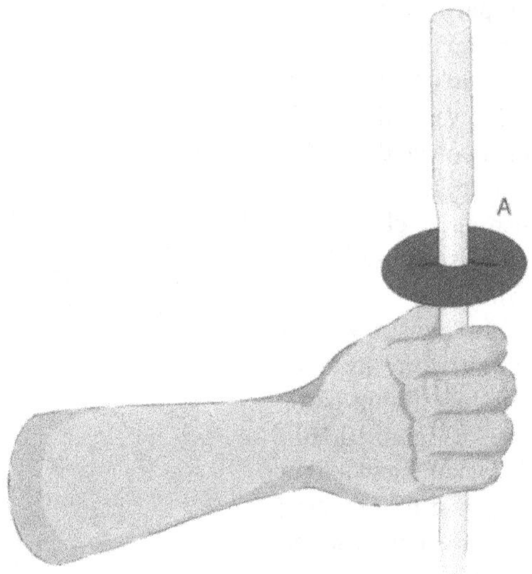

FIGURE 10.17 Total Grasp. Chisels or points should be long enough to allow total grasp by all fingers. Foam rubber (A) may be impaled on to chisel or point to protect hand from hammer strikes.

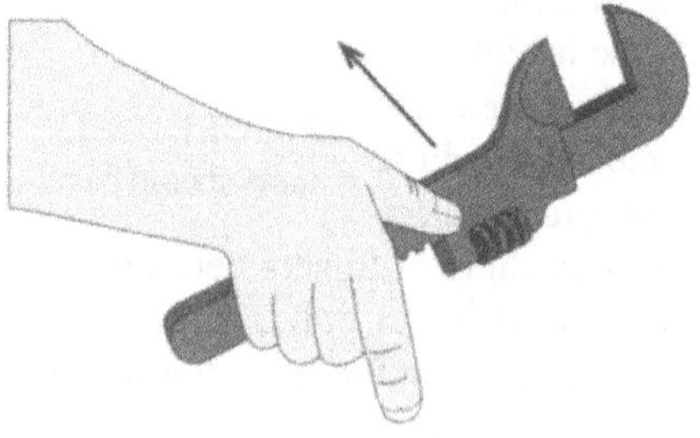

FIGURE 10.18 Pulling Wrench. Pulling wrench towards torso avoids crushing injuries to fingers if wrench slips.

FIGURE 10.19 Cheater Bar. A long pipe to increase leverage of wrench handle may snap the handle.

The operator should not hammer upon an immobile wrench handle in order to make the handle turn (Figure 10.20).

When using an adjustable monkey wrench the wrench should be placed on the nut so that the open end of the jaws face the operator.

Hardened steel tool surfaces should not be struck together. Doing so may cause hazardous flying chips.

Cutters should not be "rocked" sideways when cutting wire (Figure 10.21).

Screwdrivers should not be used as chisels or prying tools (Figure 10.22).

In order to prevent breaking or slippage, a hacksaw blade should not be installed too tightly or loosely into the saw.

Pliers should not be used on the shaft of a screwdriver to additional turning force to the screwdriver.

Sheaths, wall racks, and carrying pouches should be provided for sharp tools such as knives, axes, and hatchets.

The head of a hammer should be struck squarely and levelly upon an object ("fair blow") (Figure 10.23). The face of the hammer should be approximately three eighths of an inch larger than the object.

FIGURE 10.20 Hammering Wrench Handle. Hammering a wrench handle may break the handle.

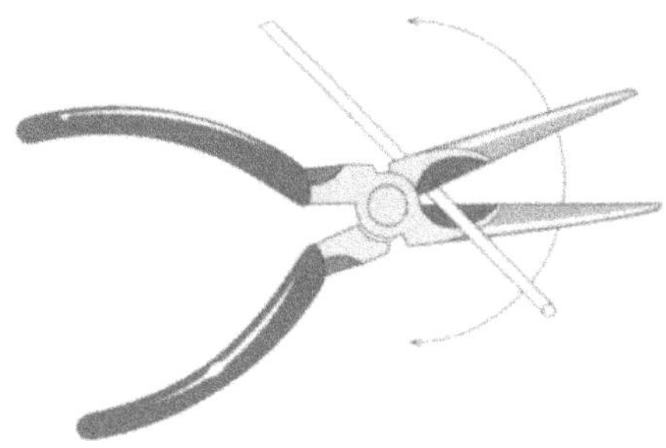

FIGURE 10.21 Rocking Wire Cutters. Rocking cutters left and right cause wire to fly out.

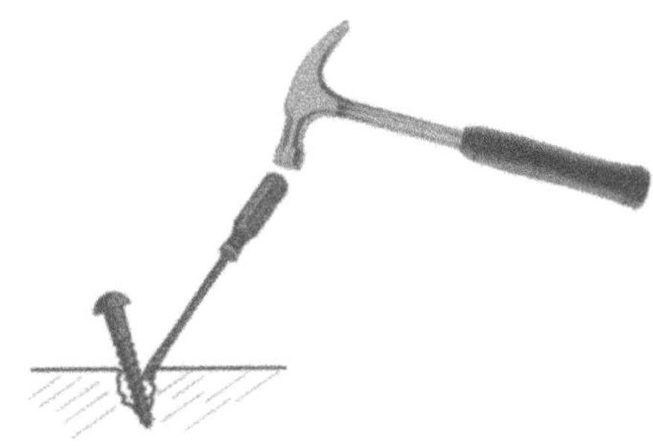

FIGURE 10.22 Improper Use of Screwdriver. Screwdrivers should not be used as chisels or pry bars.

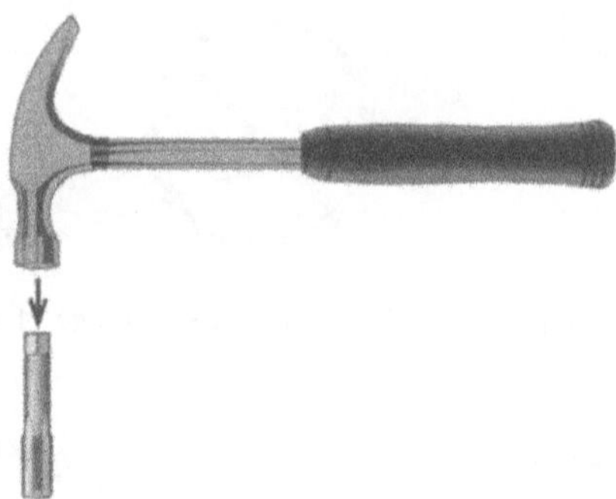

FIGURE 10.23 Fair Blow. Hammer should hit object evenly and squarely to avoid chipping hazards.

Above all: repair or discard defective tools.

Above all: use the right tool for the right job.

11 Programs

11.1 DISCUSSION

The majority of contemporary medium and large-sized organizations have established and maintain a formal safety and health program. These programs are established: 1) in response to legal directive: Occupational Safety and Health Administration (OSHA), 2) in response to complying with state of the art circumstances within their industry, and 3) in response to ethical and philosophical inclinations on the part of management.

11.2 IS A HAND TOOL SAFETY PROGRAM NEEDED?

Although overall safety and health programs have become de rigeur state of the art, hand tool programs have not prevailed in organizations. It is not suggested that such programs should always be organized. It is suggested, however, that such programs should be created if 1) there has been a large incidence of tool related injury and illness in the past and/or 2) there is a large potential for such incidence in the future. Record keeping and statistical analysis will reveal whether there have been a large disproportionate number of tool-related losses as compared to other categories of losses (remembering that tool-related incidents are generally a small percentage of overall incidents). An analysis of the organization's characteristics and industrial processes will indicate whether there is a disproportionately large use of hand tools during production. This excessive use increases probabilities of frequent tool-related accidents.

11.3 HAND TOOL PROGRAM CHARACTERISTICS

If a hand tool program is decided upon, it should be determined whether the program is a separate entity removed from other safety and health efforts or part of overall safety and health efforts. Considering relative importance and overall goals, it is advisable to write a tool safety program which is part of and included within the overall safety and health program. A tool program which is inserted into an overall safety program should contain some salient characteristics.

1. *Statement of Intent.* A program should contain an introductory section indicating: 1) management's support and commitment to the program, 2) the relative responsibilities assigned to management and workers, and 3) the goals of the program.
2. *Employee Training.* Pertinent employees should be periodically trained in: 1) selection of correct tool, 2) inspection of tools for safety hazards, and 3) safe use of tools.

3. *Periodic Inspections.* Regular inspections by supervisors or safety professionals using inspection checklists should be conducted in order to detect: 1) incorrect tool selection, 2) hazardous tools, 3) defective tools, and 4) unsafe use of tools.
4. *Tool Control.* It may be appropriate to organize a tool department if a large number and large variety of tools are needed by the organization. The tool department's major functions are 1) dispensing and keeping track of tools, 2) inspecting and maintaining tools, and 3) providing employees with advice related to tool selection and safe use.
5. *Tool Acquisitions.* There should be a cooperative effort by the purchasing department and the safety department to avoid purchasing unsafe or unhealthful tools for employee use.

12 Validation Procedures

12.1 INTRODUCTION

It may eventually be necessary to determine the effectiveness or lack of effectiveness of a specific hand tool(s) or a hand tool program which has been introduced into an organization. This may be accomplished by two research designs: 1) the before and after design and 2) the before and after design with control.

12.2 CHANGE OVER TIME

We will now consider the *before and after design*, which is described in Figure 12.1. This design belongs within the area of experimental research and involves change over time. This design is of great value to the safety and health professional because it helps him determine whether she/he is doing an effective job for his organization.

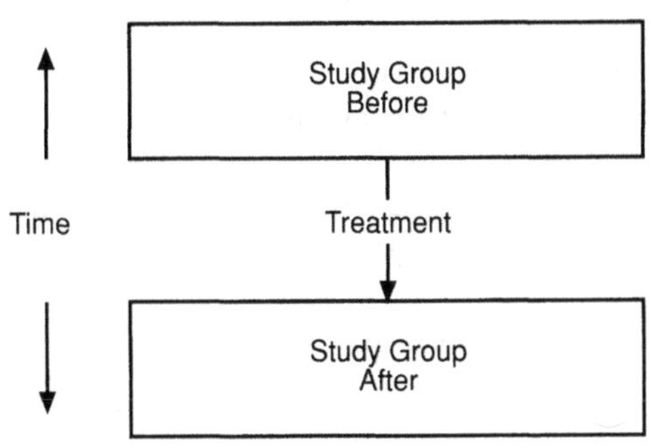

FIGURE 12.1 Before and After Design

Figure 12.2 shows how the American Society of Safety Engineers defined the "Scope and Functions" of the safety and health professional in 1966 and later in 1994 (Cacha, 1997). Figure 12.2 is as appropriate now as it was in 1966. Stated in simpler terms, the functions of a safety and health professional are to (1) discover and define safety and health problem(s) within the organization, (2) devise solutions to the problem(s), (3) communicate and implement the solution(s) within the organization, and (4) monitor the effectiveness of the solutions. Since "standing still is really nothing more than moving backwards," we can see how important it is for the safety and health professional to continuously seek beneficial changes in the organization by using the four steps in Figure 12.2.

12.3 BEFORE AND AFTER PROCEDURE

The before and after design involves the following steps:

1. Initially observing a group.
2. Affecting a change within the group.
3. Observing the group after the change.

Each activity will be discussed in turn.

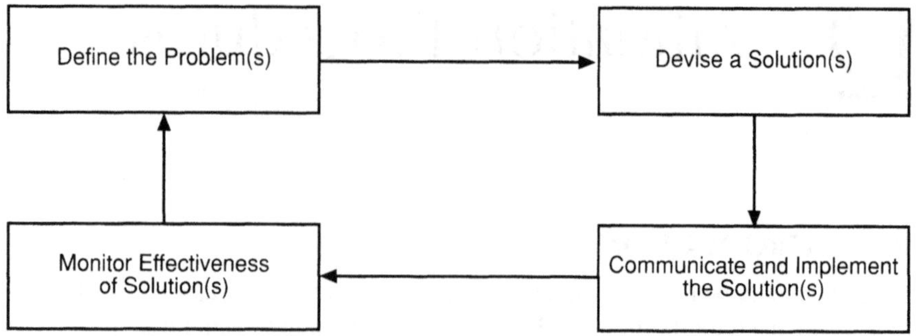

FIGURE 12.2 Scope and Functions of a Safety Professional

12.3.1 Initial Observation

The characteristics of the group which are observed must be logically and objectively chosen and must relate to the research question at hand. Observation results must be fairly, objectively, and consistently reported and recorded for future use. The observation is generally done upon one particular characteristic related to a problem area within the organization.

12.3.2 Affecting a Change

Affecting a change in a group may also be referred to as a *modification*, a *manipulation*, or, as in experimental research procedures, it may be called a *treatment*. The treatment may be of a physical nature such as guarding machines, grounding electrical equipment, or increasing illumination levels of a group's environment. The treatment may also be of a behavioral/cognitive nature such as training, motivating or informing members of the group.

Finally, the treatment may be of a managerial nature such as reprogramming, reorganizing a group, or rescheduling procedures within a group. The application of the treatment should be verifiable (are safety posters now hung in the plant?) and quantifiable (were 10 to 30, etc. posters hung in the plant?).

12.3.3 Secondary Observation

After the treatment, the group is reobserved. This secondary observation will be upon the same characteristic as was examined during the primary observation.

12.3.3.1 Prerequisites in Secondary Observation

It is important that 1) the characteristic observed in the secondary observation is the same as the characteristic in the initial observation, and 2) that the methods of analysis and observation of the secondary and initial observation are the same. As in the case of hanging safety posters, if the initial observation was the frequency of injury and illness per 100 workers, then the secondary observation should also be frequency of injury and illness per 100 workers.

12.4 CASE STUDY

Roger is a member of the safety department of the west coast plant of a large manufacturing company. Roger is presently working on his Master's Degree in Ergonomics at a local university. Because of his interest in this field, Roger's boss, the safety director, has put Roger in charge of matters related to ergonomics. Roger regularly attends meetings of the west coast plant safety committee which is composed of managers as well as representatives of labor. The committee has been recently concerned about the high incidence of carpal tunnel syndrome among employees. Some of the committee members, particularly those representing labor, are in favor of providing bent nose pliers to workers who perform frequent manual assembly. Roger is hesitant to endorse this idea because he has read that the effectiveness of such pliers has not yet been proven. Nonetheless, 1) he wishes to keep an open mind, and 2) he does not wish to give the representatives of labor the impression that the company is not interested in their welfare. Roger suggests the following experimental procedure whose implementation is supported by both upper management and the safety committee:

1. During the year 1995, Roger will discover which groups of workers are currently most exposed to carpal tunnel syndrome.
2. The number of newly-reported carpal tunnel syndrome injury cases within these groups for the year 1995 will be counted.
3. The workers within these groups will be provided with bent nose pliers and will be trained in their correct use.
4. These workers will use the pliers for one whole year (1996).
5. In early 1997, the number of newly-reported carpal tunnel injury cases within these groups during 1996 will be counted.
6. The number of incidents for 1995 will be compared statistically with the number of incidents in 1996 to determine whether the treatment (use of bent nose pliers) has been effective.

Roger begins to execute the research design. By searching the records, he has discovered that the employees working in Department 1 and Department 2 show the greatest incidence of carpal tunnel syndrome. He therefore includes only these workers in the study. He proceeds as described above with the intention of eventually comparing 1995 injuries against 1996 injuries. Roger wishes to maintain as strict a comparability between the two years as possible and, thus, wishes to take relative "busyness" of the two years into consideration. Roger discovers the following as described in Table 12.1, 12.2, and 12.3.

These tables combine before and after information on the number of carpal tunnel injuries and the payroll hours of employees for both Department 1 and Department 2. Note that Roger made an adjustment for overtime (which in this company is double time) in Table 12.2. Roger was not able to readily determine the number of employees in the before and after groups. He was, however, able to work up an equivalence between hours worked and number of employees by dividing the number of hours worked by 2,000 (the average annual number of hours worked by an employee in this company).

TABLE 12.1
Carpal Tunnel Injuries

	Dept. 1	Dept. 2	Total
1995	9	6	15
1996	5	7	12

TABLE 12.2
Payroll Hours

	Straight Time		Overtime		Total Actual Hours
	Department 1	Department 2	Department 1	Department 2	
1995	60,000	120,000	40,000	0	200,000
1996	60,000	130,000	0	0	190,000

Hours to Employee Conversion
1995: 200,000 hours/2,000 hours = 100 employees
1996: 190,000 hours/2,000 hours = 95 employees

TABLE 12.3
Injured and Noninjured Employees

	Employees Injured	Employees Not Injured	Total
1995	15 *(A)*	85 *(B)*	100 *(A+B)*
1996	12 *(C)*	83 *(D)*	95 *(C+D)*
Total	27 *(A+C)*	168 *(B+D)*	195 *(N)*

$AD = 15 \times 83 = 1245 \quad BC = 85 \times 12 = 1020$

Table 12.3 presents the information needed to perform a χ^2 test. Solving χ^2 gives a χ^2 value of only .074, which is less than the necessary 3.84 needed for the 95% level of significance. This indicates that there is no significant difference between the before and after groups, and thus, Roger concludes that the use of bent nose pliers had no effect upon the workers. The details of the solution follow:

Calculation 12.1

Chi Square (χ^2)

$$\chi^2 = \frac{N\left(|AD - BC| - \frac{N}{2}\right)^2}{(A+B)(C+D)(A+C)(B+D)}$$

$$\chi^2 = \frac{195\left(|1245 - 1020| - \frac{195}{2}\right)^2}{(100)(95)(27)(168)}$$

$$\chi^2 = \frac{195(225 - 97.5)^2}{43{,}092{,}000}$$

$$\chi^2 = \frac{195(127.5)^2}{43{,}092{,}000}$$

$$\chi^2 = \frac{3{,}169{,}968.75}{43{,}092{,}000} = .074$$

12.5 THE OTHER DESIGN

The next design, the *before and after with control technique,* is a combination of the before and after technique and a comparison technique. It is an experimental procedure similar to before and after, but is superior because it provides a control in the form of a comparison group that will provide stronger support for a researcher's final conclusions. It is a technique frequently used in educational, psychological, and social research. Figure 12.3 below describes this technique.

12.5.1 EXPLANATION

The before and after with control technique takes into consideration the possibility that the group being studied and receiving the treatment in question would have changed whether the treatment had been applied or not. This change may have been caused by various internal or external forces. For example, an exercise physiologist believes that isometric exercises are beneficial to people who are inclined to be sedentary. To test his theory, he give strength tests to a group of sedentary early adolescents. He then asks the adolescents to regularly perform isometric exercises for six months. After the six-month period, he readministers the same strength tests. He discovers that the adolescents are substantially stronger and declares that isometric exercise is beneficial for strengthening sedentary people. He reports his conclusion to one of his colleagues, who listens, thinks, and then asks, "since these are growing adolescent children, how do we know they wouldn't have gotten stronger, anyway, without the isometric exercise?" The exercise physiologist had not taken into consideration the possibility of an internal change in the adolescents in the form of maturation. His research would have been better if he had used a comparison group as well as an experimental group. The next section describes a more appropriate procedure that takes this type of unexpected (or unsuspected) change into account.

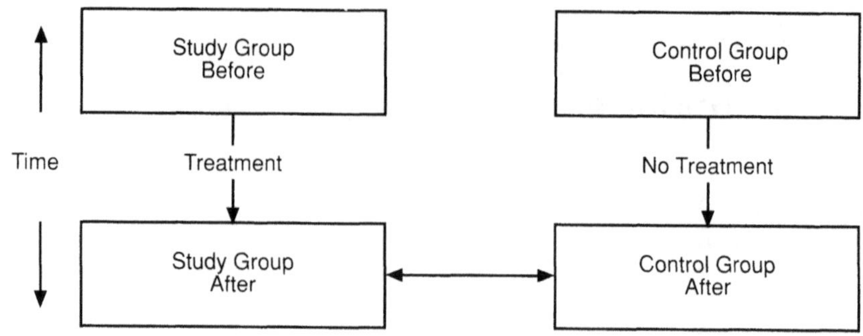

FIGURE 12.3 Before and After with Control Design

12.5.2 Before and After Technique with Control Procedure

This procedure involves the following steps:

1. Choose two groups. If at all possible the two groups should be similar in size and important characteristics.
2. Test or observe both groups with identical testing or observation procedures.
3. Apply a treatment to one group (the experimental group) but not to the other group (control group).
4. Wait an appropriate amount of time.
5. Retest or reobserve both of the groups with the same original test or observation procedure as used before.
6. For both groups, compare the original test or observation with the retest or reobservation results.
7. If there is a statistically significant change in the experimental group, but not in the control group, then the researcher's assumptions are correct and the treatment has had an effect.

12.5.3 Case Study

Ms. Weiss is the safety officer of a medium-sized corporation. The safety record for her company is above average. Nevertheless, Ms. Weiss is concerned because many employees are injured by hand tools. She therefore decides to design and implement a tool safety program. She is not totally certain that such a program will succeed, however, and she is reluctant to try it for the entire organization. The company is divided into two plants, Plant 1 on the west side of town and Plant 2 on the east side. The plants are approximately the same size and have the same operation and the same types of employees. Ms. Weiss decides to try a tool safety program in Plant 1 only. She does the following:

1. For the year 1990, she counts the number of full time employees at each plant and also counts the number of full time employees experiencing at least one tool injury at each plant.

TABLE 12.4
Off-the-Job Injuries to Full-Time Workers

	Plant 1		Plant 2	
	Total Full Time Workers	Total Full Time Workers Injured	Total Full Time Workers	Total Full Time Workers Injured
1995	506	32	490	33
1996	510	16	497	31

TABLE 12.5
Noninjured vs. Injured Workers

	Plant 1			Plant 2		
	Non Injured	Injured	Total	Non Injured	Injured	Total
1995	474	32	506	457	33	490
1996	494	16	510	466	31	497

2. At the beginning of 1996, she introduces a tool safety program in Plant 1 but not in Plant 2.
3. At the end of 1996, she collects the same type of data for each plant as she did in 1995.
4. The collected data appears in Table 12.4.

In order to perform a χ^2, Ms. Weiss changes the data as in Table 12.5. The first step is to perform a χ^2 test on Plant 2 (control group), using the data in Table 12.6.

Calculation 12.2

Chi Square (χ^2)

$$\chi^2 = \frac{N\left(|AD - BC| - \frac{N}{2}\right)^2}{(A+B)(C+D)(A+C)(B+D)}$$

$$\chi^2 = \frac{987\left(|14{,}167 - 15{,}378| - \frac{987}{2}\right)^2}{(490)(497)(923)(64)}$$

$$\chi^2 = .035$$

TABLE 12.6
Noninjured vs. Injured Workers at Plant 2

	Non Injured	Injured	Total
1995	457 *(A)*	33 *(B)*	490 *(A+B)*
1996	466 *(C)*	31 *(D)*	497 *(C+D)*
Total	923 *(A+C)*	64 *(B+D)*	987 *(N)*

$AD = 457 \times 31 = 14{,}167$ $BC = 33 \times 466 = 15{,}378$

TABLE 12.7
Noninjured vs. Injured Workers at Plant 1

	Non Injured	Injured	Total
1995	474 *(A)*	32 *(B)*	506 *(A+B)*
1996	494 *(C)*	16 *(D)*	510 *(C+D)*
Total	968 *(A+C)*	48 *(B+D)*	1,016 *(N)*

$AD = 474 \times 16 = 7{,}584$ $BC = 32 \times 494 = 15{,}808$

The next step is to compare, using χ^2, the before and after change of Plant 1 (experimental group). Using the data in Table 12.7, the calculation is as follows:

Calculation 12.3

Chi Square (χ^2)

$$\chi^2 = \frac{1016\left(|7{,}584 - 15{,}808| - \dfrac{1{,}016}{2}\right)^2}{(506)(510)(968)(48)}$$

$$\chi^2 = 5.05$$

The results of the χ^2 tests indicate a significant before and after difference for Plant 1 but not a significant before and after difference for Plant 2. Since the experimental group changed but the control group did not change, Ms. Weiss concludes that the tool safety program is an effective instrument that will provide beneficial results if applied under future circumstances.

References

Human Factors

Kantowitz, B.H. and Sorkin, R.D. (1983), *Human Factors. Understanding people — Systems Relationships*. John Wiley & Sons, U.S.

McCormick, E.J. (1964), *Human Engineering*. 2nd ed. McGraw Hill, New York.

McCormick, E.J. (1970) *Human Engineering*. 3rd ed. McGraw Hill, New York.

Sanders, M.S. and McCormick, E.J. (1993), *Human Factors in Engineering and Design*. 7th ed. McGraw Hill, New York.

Epidemiology

Cacha, C.A. (1997), *Research Design and Statistics for the Safety and Health Professional.* Van Nostrand Reinhold, New York.

Pathology

American Conference of Governmental Industrial Hygienists. (1986). *Ergonomic Interventions to Prevent Musculoskeletal Injuries in Industry.* Lewis Publishers, Clelsea, Michigan.

Naderi, B. and Ayoub, M.M. (1991), *Cumulative Hand Trauma Disorders*. Unpublished treatise presented at the University of Southern California.

Parker, K.G. and Imbus, H.R. (1992) *Cumulative Trauma Disorders. Current Issues and Ergonomic Solutions: A Systems Approach.* Lewis Publishers, Boca Raton, Florida.

Putz-Anderson, V. (1988), *Cumulative Trauma Disorders. A Manual for Musculoskeletal Diseases of the Upper Limbs.* Taylor & Francis, Philadelphia.

Anthropometry

Kroemer, K.H.E., Kroemer, H.J., and Kroemer-Elbert, K.E. (1990), *Engineering Physiology. Bases of Human Factors/Ergonomics.* 2nd ed. Van Nostrand Reinhold, New York.

Pheasant, S. (1986), *Bodyspace. Anthropometry, Ergonomics and Design.* Taylor & Francis, London.

Biomechanics

Özkaya, N. and Nordin, M. (1991), *Fundamentals of Biomechanics.* Van Nostrand Reinhold, New York.

Ergonomics

Alexander, D.C. and Pulat, B.M. (1985), *Industrial Ergonomics. A Practitioner's Guide.* Industrial Engineering & Management Press, Norcross, GA.

American Industrial Hygiene Association. (1996), *An Ergonomics Guide to Hand Tools.* AIHA, Fairfax, VA.

Chaffin, D.B. and Anderson, G.B. (1991), *Occupational Biomechanics*, 2nd ed. John Wiley & Sons, New York.

Eastman Kodak Co. (1983), *Ergonomic Design for People at Work*, Vol. I. Lifetime Learning Publications, Belmont, CA.

Eastman Kodak Co. (1986), *Ergonomic Design for People at Work*, Vol. II. Van Nostrand Reinhold, New York.

Greenberg, L. and Chaffin, D.B. (1976), *Workers and their Tools. A Guide to the Ergonomic Design of Hand Tools and Small Presses.* Rendell Publishing, Midland, MI.

McCauley Bell, P. and Crumpton, L. (1997), *A Fuzzy Linguistic Model for the Prediction of Carpal Tunnel Syndrome Risks in an Occupational Environment.* Ergonomics. 40, 8, 1997.

Mital, A. and Kilbom, A. (1992), *Design, Selection and Use of Hand Tools to Alleviate Trauma to the Upper Extremities: Part I, Part II. International Journal of Industrial Ergonomics*, Vol. 10, 1&2, 1992.

Robinson, F., Jr. and Lyon, B.K. (1994), *Ergonomic Guidelines for Hand Held Tools. Professional Safety, Journal of American Society of Safety Engineers*, 1994.

Tichauer, E.R. (1978), *The Biomechanical Basis of Ergonomics*, John Wiley & Sons, New York.

Further Reading

Human Factors

Proctor, R.W. and Van Zandt, T. (1994), *Human Factors in Simple and Complex Systems.* Allyn & Bacon, Boston.

Salvendy, G. (1987), *Handbook of Human Factors.* John Wiley & Sons, New York.

Woodson, W.E., Tillman, B., and Tillman, P.T. (1992), *Human Factors Design Handbook.* 2nd ed. McGraw Hill, New York.

Epidemiology

Spiegal, M.R. (1994), *Theory and Problems of Statistics.* McGraw Hill, New York.

Pathology

Levy, B.S. and Wegman, D.H. (1993), *Occupational Health.* Little, Brown, Boston.

Rowe, M.L. (1985), *Orthopedic Problems at Work.* Perinton Press, U.S.A.

Anatomy

Gray, H. (1977), *Anatomy, Descriptive and Surgical.* Bounty Books, New York.

Jacob, S.W. and Francone, C.A. (1974), *Structure and Function in Man.* W.B. Saunders, Philadelphia.

Van De Graaff, K.M. (1992), *Human Anatomy.* Wm. C. Brown, Dubuque, Iowa.

Anthropometry

Roebuck, J.A. Jr. (1995), *Anthropometric Methods: Designing to Fit the Human Body.* Human Factors & Ergonomics Society, Santa Monica, CA.

Kinesiology

Kelley, D.L. (1971), *Kinesiology. Fundamentals of Motion Description.* Prentice Hall, Englewood Cliffs, N.J.

Wells, K.F. and Luttgens, K. (1976), *Kinesiology. Scientific Basis of Human Movement.* W.B. Saunders, Philadelphia.

Biomechanics

Winter, D.A. (1990), *Biomechanics and Motor Control of Human Movement,* 2nd ed. John Wiley & Sons, New York.

Ergonomics

Grandjean, E. (1988), *Fitting the Task to the Man.* Taylor & Francis, London.

Mital, A. (1986), *Special Issue on Hand Tools, Human Factors, Journal of the Human Factors Society*, Vol. 28, 3, 1986.

Safety and Programming

Brauer, R.L. (1990), *Safety and Health for Engineers*, 13. Van Nostrand Reinhold, New York.

Burke, M. (1992), *Applied Ergonomics Handbook.* Lewis Publishers, Chelsea, MI.

De Reamer, R. (1980), *Modern Safety and Health Technology.* John Wiley & Sons, New York.

National Safety Council. (1988), *Accident Prevention Manual for Industrial Operations. Engineering and Technology*, 14, 9th ed. National Safety Council, Chicago.

The Hand Tools Institute. (1985), *Guide to Hand Tools.* The Hand Tools Institute, Tarrytown, NY.

Index

D

E

F

G

H

I

J

K

L

M

N

O

P

R

S

T

U

V

W

Milton Keynes UK
Ingram Content Group UK Ltd.
UKHW040051071024
449327UK00019B/465